Raven and Crow

Also by Ellen W. Williams
and from McFarland

*Imagining Heaven: Depictions of the Afterlife in World Art Across the Millennia* (2024)

# Raven and Crow

*The Mythology, Art and Science of Our Favorite Black Birds*

Ellen W. Williams

McFarland & Company, Inc., Publishers
*Jefferson, North Carolina*

Library of Congress Cataloging-in-Publication Data

Names: Williams, Ellen W., 1955– author.
Title: Raven and crow : the mythology, art and science of our favorite black birds / Ellen W. Williams.
Other titles: Mythology, art and science of our favorite black birds
Description: Jefferson, North Carolina : McFarland & Company Inc., Publishers, 2025. | Includes bibliographical references and index.
Identifiers: LCCN 2025006700 | ISBN 9781476695006 (print) ♾
ISBN 9781476655475 (ebook)
Subjects: LCSH: Ravens—Folklore. | Crows—Folklore. | Ravens. | Crows. | Ravens in art. | Crows in art. | BISAC: SCIENCE / Life Sciences / Zoology / Ornithology | SOCIAL SCIENCE / Folklore & Mythology
Classification: LCC GR735 .W55 2025 | DDC 398.24/52864—dc23/eng/20250409
LC record available at https://lccn.loc.gov/2025006700

**ISBN (print) 978-1-4766-9500-6**
**ISBN (ebook) 978-1-4766-5547-5**



Front cover photograph by Mike Hubert (Shutterstock)

Printed in the United States of America

*McFarland & Company, Inc., Publishers*
*Box 611, Jefferson, North Carolina 28640*
*www.mcfarlandpub.com*

In gratitude to the Indigenous peoples
of the world
whose myths, legends, and stories
continue to inform our lived experience
and inspire us to dream

For those with eyebrows
made of crows

# Table of Contents

# Preface

Three boisterous crows hang out in the tall oak trees surrounding my home. They cackle and tussle overhead when I walk down my driveway to the mailbox. I often pause and look up for a few minutes to see what they're doing. I can't figure it out at all. They leap from branch to branch and squawk at each other. They peer down at me every once in a while. I bow to them a few times as I pass underneath.

I'm a true nature enthusiast—but not a naturalist. I haven't got a clue what's really going on in the great outdoors and have no idea what my crows are doing up there. But I absolutely love those crows and, by extension, their crow families. They feel very familiar to me and at the same time otherworldly. Their ebony forms are mysterious, elemental, almost primeval, yet they behave like human teenagers chattering in a foreign language. I'm inclined to invent scenarios for them, to make up stories about where they have been and what adventures they are discussing. They elicit a deep psychic response in me—I think they might be magical.

Mine is not an unusual reaction. Just surf the internet and you'll discover that the crow family, Corvidae, is beloved by legions. It's the heavy metal rock star of the avian world. Scientists laud its intelligence, ornithologists study its brilliantly adaptive habits, backyard birders adore adopting it, writers are excited by its connotations, artists delight in painting it, and native peoples have built their identities around it. Of course, some people can't stand the bird. They would shudder at the sight and sounds of my crows. And some make it their mission to shoot as many as possible. There is no doubt—the bird resonates with us.

My three crows ignited in me a desire to explore the "foreign" worlds that corvids have inhabited. My initial stop was the world of myth, legend, and fairytale. I have always been captivated by the ancient stories of Raven, a trickster-creator figure of the Arctic North. In circumpolar mythologies, Raven is frighteningly powerful but also sometimes a silly and amusing fellow. As I rummaged and rooted through crumbling library books, I soon discovered that crows and ravens feature in mythologies from every corner of the planet! Thousands of tales exist to describe the activities and attributes of the supernatural bird. Some are light-hearted, others are cautionary lessons. They all speak to the deep connection we have developed with this species.

My next stop was the world of art. What have artists done with the bird? Again, the range of work is immense. Traditional bird artists render the crow or raven so that we can see the fine details of its features. Contemporary painters highlight its more suggestive and enigmatic qualities. Poets and writers channel its human-oriented mysteries. Through the artist's eye, we find a bird that is not ordinary. It is beautiful, and redolent with meaning. But what of its true nature? That too turned out to be an exploration of a mysterious world. Scientists are continually amazed by the capacities of corvids. New findings reveal a use of "language," problem-solving skills, an ability to count, and a "theory of mind" whereby the bird can understand the mental state of another living being and use this information to find food or escape danger.

Taken together, we have a portrait of an amazing bird. The following chapters are intended to construct this portrait. Because myth represents our most basic spiritual connection to the natural world, myths and tales are used to form the core of the material in this book. Around the stories are nestled aspects of science, art, and ethnology. Looked at from these many angles, crows and ravens emerge as truly awesome creatures who inspire us to wonder, to examine, and to celebrate. We can find ourselves in one of nature's wild children.

Many people have helped to bring this project to life: among them, readers Andy Kroll and Pam Hart, supreme counselor Sarah Schultz, and cheerleaders Jim Fusilli and Adrienne Kirschenbaum. Most importantly, however, an acknowledgment of gratitude is due to the Indigenous peoples whose stories are used in this collection. This work is dedicated to those who have worked so hard to protect, preserve, and share their heritage. We are all truly enriched by their efforts.

# Introduction

A black speck drifts just above the tree line. It is moving on a warm thermal breeze. The sight of it might stop you in your tracks and, with squinting eyes, you might follow it with wistful curiosity. What bird *is* that: a crow or a raven? Most of us can't tell them apart, so here are a few tips:

- Ravens are larger than crows. A raven weighs about two and a half pounds and has a wingspan of around four feet. A crow weighs in at just about one pound and has a wingspan of around three feet.
- Ravens tend to soar; crows flap more often.
- A raven's beak is big and thick with a distinctive downward hook shape while a crow carries a more delicate, narrow, pointed bill.
- A raven's tail is wedge-shaped; a crow's tail parts into a fan.
- A raven will "quork" at you while a crow is more likely to "caw."

Crows and ravens belong to the order *Passeriformes* (which are songbirds) and to the genus *Corvus*. The common raven is known as *Corvus corax* while the American crow is a *Corvus brachyrhynchos. Corvus corax* is the world's largest perching bird and North America's largest songbird. Both birds share the genus *Corvus* with over forty other bird members, such as the white-necked raven of Africa, the slender-billed crow of Malaysia, the fan-tailed raven of the Middle East, and the Jamaican crow. Crows and ravens are also related to jays, magpies, treepies, jackdaws, and nutcrackers.

Behind the glassy black eyes of both ravens and crows lies a brain that amazes every scientist who studies it. Dr. Lawrence Kilham said: "Crows, and with them I include ravens, seem as though by convergent evolution to have something in their psyches corresponding to something in our own."[1] An enlarged forebrain equips the birds with an ability to think in flexible, nuanced ways. They can consider, reconsider, and adjust their behaviors. They can remember the past and think about the future. This species of bird has been able to adapt to every environment on the planet—deserts, forests, mountains, and tundra. The same species of raven found in the Sahara is also found in the Arctic.

Although the birds are objectively "plain," they have a way of beguiling us. We are irresistibly drawn to their blackness, to their cleverness, to their strangeness. Watch them for any length of time and it will be hard not to think of them in human terms. They seem more like people than people. Charles Dickens once described a raven as a "very particular gentleman with exceedingly tight boots on, trying to walk fast over loose pebbles."[2] The image is unforgettable. The Rev. Henry Ward Beecher is said to have commented: "If human beings wore wings and feathers, very few of them would be clever enough to be crows."[3] That just may be true. Even in these jaded, contemporary times crows and ravens are magnetic. Think of the band Counting Crows, or Lou Reed's 2003 album *The Raven*, or the Royal Ballet of London's 2013 adaptation of the graphic novel *Raven Girl*, or the three-eyed crow in the television drama *Game of Thrones*.

Because they are so captivating, crows and ravens have been given outsized roles in our cultural history. We have lived with them for tens of thousands of years and have learned to think through their movements and activities. The birds are so present in our lives as to cross the boundaries of reality into unseen worlds recounted through myths and stories. Here we have one of the world's great characters, and the words and images left from lifetimes ago reveal our long-standing dialogue with it.

The first depiction of a crow or a raven is believed to be in a prehistoric cave painting in Lascaux, France. Crow and raven images have also been found on ancient Celtic fleshforks, Mithraic tauroctonies, Egyptian friezes, and Japanese coins. To the philosophers of Persia and India, the raven was divine, of celestial origin and supernatural ability. In Japan, Taoist lore describes a three-legged crow inhabiting the sun. Shinto beliefs see the crow's role as that of messenger to the sun goddess Amaterasu. Korean Buddhist art of the sixth century CE shows the Sunlight Bodhisattva wearing a sundisk with a crow inside. Indeed, the crow and raven have an almost universal association with solar deities and symbols.

Throughout the world, the raven and crow have been used for many forms of augury. In Tibet, an ancient ninety-square chart interpreted the meaning of one or another sound coming from a raven's voice. In India, the cawing of a crow meant the arrival of a guest or the return of a beloved one, and an elegant sculpture from the ninth century CE found in eastern India depicts a wife waiting at a half-open door with a crow on her shoulder. In Europe, the mere presence of a crow or raven brought omens. If you see crows flying toward you in England, beware: "One's unlucky, Two's luck, Three is health, Four is wealth, Five is sickness, And Six is death." In fact, folklorist Ernest Ingersoll once wrote that "no bird, save possibly the cuckoo, is so laden with legends and superstitious veneration as the raven."[4]

Their usefulness as bearers of omens further associates ravens and crows with the shamanic world. Anthropologist Mircea Eliade noted that ravens were perceived to be the guardian spirits of shamans (practitioners of divination and healing). Across the Northern Hemisphere, the birds were conscripted to amplify shamanic powers.[5] Among the Chukchee of Siberia, the chief deity Big-Raven (*Kuyginn-o'qu*) had a raven assistant (*Ku'urkil*) who was called upon to cure patients. Ku'urkil was able to devour disease the way a bird devours worms. Elsewhere in the north, the well-known raven rattle, used extensively by shamans of the Northwest Coast, summoned the healing spirits and drove away evil forces.

As navigators and guides, the crow and raven are renowned. The Inuit peoples assert that ravens guide hunters toward game. The Greek writer Pliny (first century CE) corroborates, reporting that a certain Crates Monoceros at one time hunted with the aid of ravens "to such an extent that he used to carry them down into the forest perched on the crest of his helmet or on his shoulders; the birds used to track out and drive the game, the practice being carried to such a point that even wild ravens followed him in this way when he left the forest."[6] The Emperor Jimmu of Japan (seventh century CE) marched under the guidance of a gold-colored raven and Alexander the Great (fourth century BCE) was said to have been led across the desert from the Mediterranean Coast to the oasis of Ammon by two ravens sent from heaven. The Egyptian pharaoh Mares (Amenhotep III), fourteenth century BCE, reportedly owned a wonderfully special crow that he used as an express delivery service for dispatches. The crow never failed to find its destination and return home. Mares honored the crow's services at death with its own tomb and monument.[7] Finally, the Viking "Saga of Floki" relates how in 864 CE Floki, a legendary explorer, successfully took three ravens with him as guides.

Ravens and crows do not merely have supernatural attributes. These birds are supernatural beings. In India, once a year the deceased assume the bodies of crows to whom food offerings must be made and accepted. The British god Lug was a raven, while the Nordic deity Odin was called *Rafnagud* or raven-god. In Russia, crows were spirits of witches. In Germany, ravens were damned souls. Raven is the supreme creator among Northwest Coast Indian tribes such as the Tlingit and the Haida as well as among Siberian tribes such as the Koryak and Chukchee.

In more modern times, we find crows and ravens closely associated with saints, Christianity's near equivalent to supernatural beings. St. Vincent's body was rescued by a raven and borne to his brothers at Valencia. It was said his tomb was constantly guarded by faithful ravens. St. Meinrad apparently lived alone in a cell with two ravens. He was murdered by two men who fled, only to be followed and harassed by the ravens until they surrendered. "The Miracle of the Raven," pictured in a fresco in Florence, depicts St. Benedict as his life is saved when a raven removes poisoned bread from his table. We even have the celebrated story of St. Anthony and the hermit Paul who were fed by ravens. Of course, not everyone in Europe felt that the birds belonged with saints. French peasants used to say that bad priests became ravens and bad nuns became crows. St. Ambrose, in a treatise, devotes a whole chapter to the raven's impiety in not returning to the ark.

Mythology and storytelling have long served the essential function of passing on important cultural ideas from generation to generation. Scholar Joseph Campbell adds that an additional purpose of mythology "is to waken and maintain participation in the mystery of this finally inscrutable universe."[8] Mythologies help us to live—they lay out the rules of society, explain why things are the way they are, keep the past present, and reveal the unseen nature of reality. Somehow, in our rush toward the future, we have become disconnected from the mythological past. Although you might encounter phrases such as "Go to the crow!" for drop dead or "You have a raven's knowledge" to praise the ability to find a lost object, our stories are thin and dwindling. And our imaginations suffer for it. Our creative understanding of life is certainly diminished. The crows and ravens we see around us offer everything from wisdom and practical advice to entertainment and raucous humor. They are perfect avatars for meaningful narratives because they embody the random, capricious ways of nature while simultaneously showing us how to negotiate it all. We might benefit by reconnecting with them.

Luckily, scientists and journalists seem to have stepped in to occupy the space left bare of the mythmakers. Recent research confirms truths behind many legends and the factual stories now told about the birds in their natural surroundings enthrall readers anew. A steady stream of popular science articles, published every year, highlight new discoveries being made about the birds. Crows can understand water displacement theory. Ravens are smart enough to be paranoid. Responding to the Stanford University "marshmallow experiment" to test delayed gratification in children, crows will exhibit self-control to wait for a better treat later. Ravens will recognize valuable items and put them aside for future use. Corvids are now believed to stand at the pinnacle of avian species. The lowly crow and raven have attained the status of "super-bird" among those who truly know them. This entirely new set of stories, generated by researchers, can help our efforts to connect anew with the birds.

But, to journey back to the old stories and travel around the world to revisit ancient tales and artifacts is to cover an entire history of our dialogue with this particular animal species. Through myth, legend, and image, Raven and Crow are able to reveal themselves in their many forms and disguises. We see in these enigmatic figures a magical clown, a benevolent mischief-maker, and an omnipotent spiritual power who has engineered, altered, and manipulated the world to its present form.

This book looks at how science, art, and storytelling interrogate the mystery of a simple black bird. Presented here are the voices of those who have paid attention to the crow and the raven. It is worth pausing at each piece to consider what precisely is being shared or offered to us. What can we take away—about ourselves, about the bird, or about the world at large? The pieces included in this collection come from numerous sources. They have been grouped into ideas that revealed themselves as research progressed. In each group, however, multiple themes resonate—within a single story and between stories. The Inuit story of the raven and the loon, for example, is as much about beginnings and interspecies relations as it is about blackness. The Arapaho Ghost Dance songs are about life and death, sky associations, and shamanism as well as about avian messengers. Throughout the book, recurring motifs from vastly different peoples over hundreds of years reverberate between one another to tell us a bigger story about humankind's relationship with the crow and raven.

Every collection of translated material presents the reader with the problem of identifying authorship. This is especially difficult when looking at historical documents. In all cases here, the most original source has been sought for each story. In most cases, however, a non-native translated the story into English and sometimes the translation went through more than one language to English. Although it is now the custom for a folklorist to bring a translation back to the original teller for approval, this has not routinely been the practice. Since translators inevitably bring their own cultural biases to their work, the stories must be read with that added textual layer in mind. Therefore, in order to provide a more complete rendering, the source and cultural context of each story has been identified, as well as the name and background of the collector-translator where relevant. Wherever possible, Indigenous names and identifications are presented in two ways—as the anthropologist noted them and as the people identified themselves. Providing this additional background serves to enrich the piece itself. The reader can envision the storyteller and the audience as well as understand who was committing the narrative to paper.

The terms myth, legend, fairytale, folklore, and story each have different definitions which can sometimes be useful for a Westerner, but which are often meaningless to a traditional teller who does not make similar distinctions. As a rule, a myth is believed to be true and sacred. Commonly set outside of conventional time, it usually explains how and why things came to be the way they are. A legend is rooted in an identifiable place, time, or person. It feels possible but maybe not entirely true. A fairytale deals with enchantment, fantasy, and the magical but frequently has a moral point. Folklore comprises all the oral traditions of a group of people, from simple sayings to elaborate myths. And then there are stories—epic adventure stories, tragedies, parables, tales, fables—all of which present us with rich, enlightening worlds that intersect our own.

Anthropologists and folklorists today are reexamining the content and legacy of their work. Early social scientists approached their subjects with mindsets and paradigms that we would object to today. They created frameworks that aided their own purposes without much regard for the needs or approval of the peoples they rested their careers on. Nevertheless, these researchers, while representatives of the dominant white society, were for the most part genuinely interested in learning from the Indigenous peoples they visited. They believed that native heritage had something essential to offer future generations. While we must acknowledge that the stories were not created for us, we know that without them we would all be poorer in spirit.

The crow and raven are featured to the exclusion of other members of the species, because they are so similar in appearance. Scientists as well as storytellers lump them together. They are, for most purposes, the same bird—one smaller, one larger. A bird that still holds secrets.

Much is not understood about the natural history of the crow and raven. Researchers have many more hypotheses to test about their cognitive abilities. And much remains undiscovered among the many myths and tales that lie buried in minds of the elders, or among the millions of pieces of digital data, or inside the dusty old books I visited that languish in libraries. It is my hope that in reading this collection of fact, image, and story the reader will come to understand the significance crow and raven have held in our collective imagination. We have epic protagonists in our midst. The stories they have provoked, both scientific and mythic, help us make sense of our place in the world. We all live our lives as narratives—as stories we tell each other about ourselves. Here are the stories we have told. And here are Crow and Raven—true originals who have been with us since the dawn of humankind.

## 1

# Living in Black and White

When I walk out from my garage doors at the first light of day, I know that when I glance up I will see three motionless silhouettes above me. Round-shouldered bodies will be sitting slightly hunched on a tree branch, heads tucked in. My crows often look fake. Depending on my mood, I sometimes think they look like shooting range cutouts while at other times I shiver at their bleak countenance. The raven and the crow have the distinction of being among the darkest of bird species. Their sooty, monotone pelts bestow upon them a kind of mysterious power; blackness seems to speak to our darker side. In observing a crow or raven that is stationary, one is almost affronted by its bold outline. With its form alone, it creates negative space. The image is startling, disturbing, and insistent upon an answer to a question we haven't even contemplated yet. Edgar Allan Poe certainly felt this way. Writer Cao Wenxuan writes: "… that black is truly black! As black as ink, as black as lacquer, as black as the dark night unlit by moon and stars. Yet, there is a sheen to it, and when the crow takes flight, it shimmers like satin in sunlight."[1] And, Geologist Rick Bass expressed some of our feelings when he wrote: "Ravens, black as coal, shiny and greasy, flying in sun, like winged, black devils…. I feel as if I'm on their side, and it scares me,…."[2] Their silhouettes beg for a story.

## *Field Notes: Black Feathers*

One might presume that all-black feathers would be a problem for the birds, making them both ill-suited to hot climates on the one hand and too noticeable in snow-bound territories on the other. In fact, the dark plumage works quite well. First, black feathers are strong feathers. The melanin pigment that colors them black binds to the keratin protein in the feathers to make them stiff, durable, water and wear-resistant—perfect for traveling long distances in challenging climates. Second, the jet-black feathers serve to concentrate heat from the desert sun at the feathers' surface rather than closer to the skin. These tough feathers protect inner organs from ultraviolet radiation damage while a sudden breeze will move the heat away quickly, preventing overheating. In cold environments it works the other way. The pelt absorbs a lot of solar radiation on a sunny day, allowing the birds to conserve their body heat (that is, if they stay out of the wind …). In forested areas, their figures will blend perfectly into the long shadows of ancient trees should they wish to hide or creep up on their prey. In the night, their blackness helps to make them invisible to the predators that seek them out. Conversely, in daylight, if they want to make themselves known quickly to potential mates or teammates all they have to do is step out into the open.

Because they cannot take advantage of bright plumage for broadcasting their intentions to one another, corvids signal in other ways. They will flash the whites of their eyes if angry and flick a tail that would be seen from a great distance. Or, they can use any one of a number of calls. The birds can also posture with their bodies. A subtle arrangement of feathers will express a specific emotion and researchers can tell exactly how a crow feels just by looking at how it has positioned its body. So can all the other crows around it. When afraid, a raven will will press back back its head feathers; when courting, it will fluff them up for a maximum degree of bravado.

The sleek profiles of the raven and crow impress us with their simplicity. From a distance they appear to us as unknowable specters—dark beings who have made their home with us. We can't stop wondering what they are up to. And we can't stop giving voice to the stories of their lives.

## How the Crow Came to Be Black

Brulé Sioux (Sičhą́ǧu Oyáte) legend. Told by Good White Buffalo at Winner, Rosebud Indian Reservation, South Dakota, 1964. Recorded by Richard Erdoes.

In days long past, when the earth and the people on it were still young, all crows were white as snow. In those ancient times the people had neither horses nor firearms nor weapons of iron. Yet they depended upon the buffalo hunt to give them enough food to survive.

Hunting the big buffalo on foot with stone-tipped weapons was hard, uncertain, and dangerous. The crows made things even more difficult for the hunters, because they were friends of the buffalo. Soaring high over the prairie, they could see everything that was going on. Whenever they spied hunters approaching a buffalo herd, they would fly to their friends and, perching between their horns, warned them: "Caw, caw, caw, cousins, hunters are coming. They are creeping up through that gully over there. They are coming up behind that hill. Watch out! Caw, caw, caw!" Hearing this, the buffalo would stampede, and the people starved.

The people held a council to decide what to do. Now, among the crows was a huge one, twice as big as all the other crows. This crow was their leader. One wise old chief got up and made this suggestion: "We must capture the big white crow," he said, "and teach him a lesson. It is either that or go hungry."

He brought out a large buffalo skin, with the head and horns still attached. He put it on the back of a young brave, saying: "Nephew, sneak among the buffalo. They will think you are one of them and you will capture the big white crow."

Disguised as a buffalo, the young man crept among the herd as if he were grazing. The big, shaggy beasts paid him no attention.

Then the hunters marched out from their camp after him, their bows at the ready. As they approached the herd, the crows came flying, as usual, warning the buffalo: "Caw, caw, caw, cousins, the hunters are coming to kill you. Watch out for their arrows. Caw, caw, caw!" and as usual, all the buffalo stampeded off and away—all, that is, except the young hunter in disguise under his shaggy skin, who pretended to go on grazing as before.

Then the big white crow came gliding down, perched on the hunter's shoulders, and flapping his wings, said, "Caw, caw, caw, brother, are you deaf? The hunters are close by, just over the hill. Save yourself!"

But the young brave reached out from under the buffalo skin and grabbed the crow by its legs.

With a rawhide string he tied the big bird's feet and fastened the other end to a stone. No matter how the crow struggled, he could not escape.

Again the people sat in council. "What shall we do with this bad, big crow, who has made us go hungry again and again?"

"I will burn it up!" answered one angry hunter, and before anybody could stop him, he yanked the crow from the hands of his captor and thrust it into the council fire, string, stone and all. "This will teach you," he said.

Of course, the string that held the stone burned through almost at once, and the big crow managed to fly out of the fire. But he was badly singed, and some of his feathers were charred. Though he was still big, was no longer white.

"Caw, caw, caw," he cried, flying away as quickly as he could, "I'll never do it again; I'll stop warning the buffalo, and so will the Crow Nation. I promise. Caw, caw, caw."

Thus the crow escaped. But ever since that time, all crows are black.

This story has been recounted hundreds of times in many iterations, probably because it is both entertaining and educational. At the start of the story impressive white crows purposefully work in coordination and on behalf of another animal species. Nature is protecting itself. By the end of the story we see a shift toward humans. We witness an ecosystem working itself out through the triangulation between bird, beast, and human as they negotiate for resources. We are in the beginning of things after all, and humankind must defend its place.

The Brulé are a branch of the Lakota Sioux people now living primarily in southwestern South Dakota. When white European settlers arrived in the seventeenth century, they encountered the Lakota in and around present-day Minnesota. By the 1700s, however, these expert horsemen, buffalo hunters, and warriors had expanded their territory to include the high plains of Wisconsin, Iowa, the Dakotas and parts of Canada. After the battle at Little Bighorn on June 25, 1876, the Sioux were separated into their various groups. Today, the Rosebud Sioux tribal people struggle, but continue to practice their traditional ways on the Rosebud Indian Reservation.

## The Raven and the Loon

Told by Arnarquik, Inuit, in Nahigtartorvik, Kazan River, Nunavut, Canada. Collected and translated by Knud Rasmussen.

In the olden days, all birds were white. And then one day the raven and the loon fell to drawing patterns on each other's feathers. The raven began, and when it had finished, the loon was so displeased with the pattern that it spat all over the raven and made it black all over. And since that day all ravens have been black. But the raven was so angry that it fell upon the loon and beat it so about the legs that it could hardly walk. And that is why the loon is such an awkward creature on land.

The importance of Raven in Inuit and northern cultures is hard to overestimate. From Alaska to Greenland, this bird is a central figure. Raven stories have circled round fire pits and filled the nighttime air for generations. In the north, Raven is everywhere—as the Creator, the trouble-maker, the glutton, and the hero. He is responsible for game, the weather, light itself, good and bad luck, fertility, and death. In Alaska, the heavenly galaxy is called "the trail of the Raven's snow-shoes."[3]

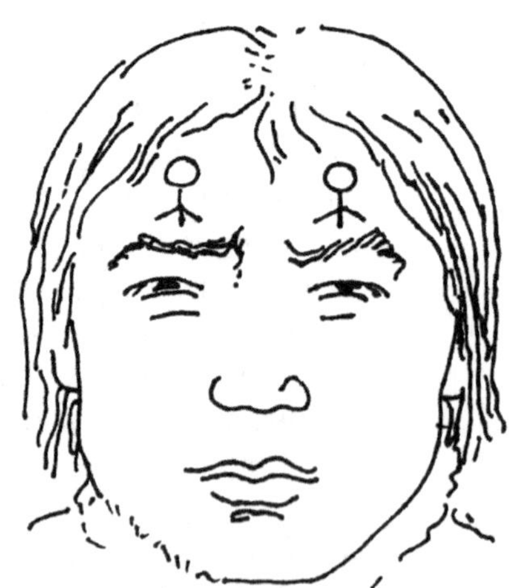

**Raven tattoo. ca. 1900. Redrawn by author. Alaska. Illustration from E.W. Nelson, *The Eskimo about Bering Strait*. The practice of tattooing is centuries old in the Arctic North. Symbols etched on the body were powerful statements of identity as well as talismans for good luck. The tradition is very much alive today and highly valued among indigenous peoples.**

Here, Raven is both inept and temperamental; he cannot draw well and he is vengeful. In another version of this story, told in the 1950s by Thomas Kusugaq at Repulse Bay, Nunavut, the raven and loon set about tattooing each other. Raven, however, is unable to stand the painful procedure and jumps about. Loon gets aggravated and takes the soot drip pot from an oil lamp and dumps it onto Raven. Before Loon can flee, Raven throws the pot at him, hitting his legs. According to Kusugaq: "Because of these occurrences, loons can never walk and, because the loon poured the drip pot over the raven, ravens are now black."[4] A Greenlandic version of this same story also exists.

**Studio portrait of Knud Rasmussen (right) and Danish photographer Leo Hansen (left), with Miteq and Arnarlunnnguaq, ca.1924. Lomen Bros. photographer. Alaska. Rasmussen and his two Inuit companions completed a great sledge journey across the Arctic North together. Hansen joined as staff photographer on the final stage of the expedition (University of Washington Libraries, Special Collections, UW41957. PH Coll 413.Album 14.1 Ralph E. MacKay Alaska photograph albums).**

A significant feature of the Arctic raven complex (the stories, shamanic practices, and rituals involving Raven) is his representation as marks on objects and as tattoos on bodies. The symbol of the raven links the bird to a web of cultural and mythic history. Our story hints at the rich Inuit tradition of tattooing (one that continues even today) which was widespread, intricate, and prized. Tattoos signal tribal affiliation, personhood, and religion. They also channel the world of the spirits and are thought to provide protection from enemies. In some areas, it was believed that only tattooed women would be allowed to enter the pleasant afterlife in the sky. Alethea Arnaquq-Baril, an Indigenous woman living today, received hand and facial tattoos, explaining to writer and photographer Lars Krutak:

> "Many elders told me that 'your tattoos stay with your soul after you die,' so when you pass on you can recognize your family based on the tattoos they [carried] into the afterlife." And because tattoos were added to commemorate life accomplishments like killing your first

> animal or making your first parka, she added "If you're handsomely tattooed, then obviously you have done many good things in your life and you have a high tolerance to pain, and also an inner strength."[5]

Among specific groups, the three-forked figure was a raven totem (emblem for a specific family, clan, or tribe). Babies were sometimes wrapped in raven skins to safeguard them. And the raven sign, painted on objects such as smoke-hole covers or implements, augured future success at the hunt.

Knud Rasmussen (1879–1933) was a polar explorer and anthropologist who lived intimately with the peoples he yearned to know more about. He travelled for nearly thirty years by dogsled throughout the polar world collecting detailed information about geography, Greenlandic and Inuit material culture, and also spiritual beliefs and ritual culture. In November 1921, his most ambitious Fifth Thule Expedition commenced. After 18 months exploring the enormous terrain west of Hudson Bay, Rasmussen and two Inuit set off on what would become the longest sledge journey in history and the first through the entire Northwest Passage. They traveled 20,000 miles over three-and-one-half years, traversing Canada and Alaska to Point Barrow, the most northerly permanent settlement in North America at the time. They even reached Siberia for a brief stay before being turned away by Soviet authorities. Rasmussen's account of his expedition, published in English in twenty-six reports but compiled in the 1927 book *Across Arctic America*, is considered a classic of polar expedition literature.

## Wohpekumeu and Crow

Told by Julia Wilson and Kate (her mother), Yurok (Oohl) in Wahsek, on the Klamath River, northwestern California. Collected and translated by Alfred L. Kroeber.

At first people were going to eat pitch because they had no acorns. Wohpekumeu was the one who made everything along this river. He went to the sky and saw people eating acorns. He hid one in his mouth. The people saw and pursued him. Fleeing, he ran into a hollow tree and it closed on him. He could not come out.

A bird came and pecked on the tree. Then Wohpekumeu told it to bring more birds and free him. It brought others and they all worked (at pecking a hole). Crow came too. At last Wohpekumeu got out.

Now he said that (in return) he would paint them pretty. Crow said, "I want to be red woodpecker crest all over my body." Wohpekumeu said, "I will make you so but you will then have to stay far back in the woods." Crow said, "I do not want to be back in the woods. I want to be about the towns, and yet I want to be hard to catch." Then Wohpekumeu became angry, for he wanted to make the birds so that they could be used by people for woodpecker crest headbands. So he said to Crow,

"I will make you." Now he told all the birds to close their eyes while he painted them and then to fly off to wherever they liked, and there look at themselves. Then he painted Crow, but black all over.

Crow flew off to a town. When he looked at himself and found himself wholly black, he was angry, for he had worked the hardest to free Wohpekumeu. He flew back to the place where they had been painted and found a few little birds still around. He shoved these into the ashes, and that is why they are dull and not so pretty as the others.

Crow is vain; and he is picky. He wants it all—to be colorful and flashy but not a target for hunters; to be noticed by those in town but not a subject of their interest. It does him no good when faced with Wohpekumeu, the complicated culture hero of the Yurok people who was a seducer, womanizer, and gambler but also the one responsible for stealing and making available life-giving acorns and salmon, for directing the course of the Klamath river, for showing the sun how to travel across the sky, for distributing medicine, and for easing childbirth. It was said about Wohpekumeu that "he knows every kind of play. He can make any kind of food grow. He can make anything be that he wants. He thinks of it and it grows."[6] Crow could only ever hope to be his sidekick.

Alfred L. Kroeber (1876–1960) collected Yurok stories between 1900 and 1907 as one of America's earliest anthropologists. His initial interest was in linguistic geography but he developed a life-long fascination with the Yurok, characterizing them as "a remarkable civilization." He recorded many versions of this story. In his remarks he stated that he was "pleased to record from the mouths of women." Julia Wilson was "trained in the manner of the whites" but her mother was "a full-blooded Yurok living in an ancient Indian town."[7] Kroeber is seen in a less than favorable light today because he demonstrably "othered" his informants in his work and held the superior attitude that was common at the time. Nevertheless, his collection of myths and stories were crucial in preserving the oral heritage of a culture that was quickly dissolving.

## Wak and the Raven

Told by Professor Ahmed Mohammed Ali, Somalian Ethiopia.
Collected and retold by Elizabeth Laird with help from Mikhail Negussie.

Once upon a time, there was a Sun God. His name was Wak. The Raven was Wak's messenger. He used to fly from the earth to the sky and from the sky to the earth with messages in his beak. In those days, the Raven was a great sheikh. He was all white, and there was not a black feather on him.

Now the birds could not live together peacefully. They fought every day over their food. The little birds tried to take the meat from the big birds. The big birds tried to steal the seeds and fruit from the little birds. They could not agree. At last, they called a meeting.

"We must share our food between us," they said, "or there will never be peace. We need a judge to help us. Let's ask our sheikh, the Raven. He can tell us what to do."

The Raven listened carefully to all the other birds. "I will ask our master, Wak, the Sun God," he said, and he flew up into the sky. The other birds waited for the Raven for a long time. At last, he flew back to them.

"Listen," he said. "Wak has given me the answer to your question. Let the big birds eat only meat. Let the small birds eat only fruit and seeds."

"But who is a big bird? Am I big?" asked the partridge.

"And who is a small bird? Am I small?" asked the parrot.

The Raven thought for a moment.

"If the birds are bigger than me, they are big birds," he said. "They can eat meat. But if they are smaller than me, they can only eat fruit and seeds."

"And you?" a little bird called out. "What will you eat, Raven?"

The Raven smiled.

"Oh," he said, "I am in the middle, not big, not small. Also, I am your sheikh and the messenger of Wak. I can eat meat and fruit, and anything I like."

The other birds were angry.

"You have cheated us!" they said. They tried to catch the Raven, but he flew away from them. He went up, up into the sky, to find his master, Wak.

"Wak!" he called. "Wak! Wak! Help me!"

On he flew, closer and closer to the sun. The heat burned his feathers, and they became black. Now there was not a white feather on him.

From that day, the Raven's feathers have always been black, and he still calls out, "Wak! Wak! Help me!"

Although much of the material included in this book comes from the distant past, oral literature is still being created today. Storytelling is still valued, in spite of our modern times. Elizabeth Laird, who had earlier lived and worked in Ethiopia, was commissioned by the British Council and the Ethiopian Ministry of Education to travel across its landscape of dizzying diversity to collect the legends and folklore that are told by everyday people—priests, farmers, merchants, midwives, and children.

From a politically dangerous Ethiopian region of great beauty comes the Sun God Wak, known by both the Oromo and Somali people. Wak was the supreme god before the Christian and Muslim God. He oversaw a land where camels trod through grasslands and villagers rested under enormous sacred trees. His messenger and chief scout, Raven, comes off as a crafty creature in the story. Although he has a special position in relation to the highest deity, Raven still seems concerned about protecting his own interests. Keep this Raven in mind when reading about Odin's ravens, who have similar roles as scouts and messengers, and Raven of the Arctic north, who also gets into trouble for selfishness. Wak did not intervene to save his messenger—we might draw a lesson from this fact.

## The Crane and the Crow

Wiradjari, New South Wales, Australia. Told to K. Langloh Parker.

The crane was a great fisherman. He used to hunt out the fish, with his feet, from underneath the logs in the creek, and so catch numbers.

One day when he had a great many on the bank of the creek, a crow, who was white at that time, came up. He asked the crane to give him some fish.

"Wait a while," said the crane, "until they are cooked."

But the crow was hungry and impatient, and would not cease bothering the crane, who kept saying, "Wait. Wait."

Presently the crane turned his back. The crow sneaked up and was just going to steal a fish. The crane turned round, saw him, seized the fish, and hit the crow right across the eyes with it. The crow felt blinded for a few minutes. He fell on the burnt black grass round the fire, and rolled over and over in his pain. When he got up to go away his eyes were white, and the rest of him black, as crows have been ever since.

The crow was determined to pay out the crane for having given him white eyes and a black skin.

So he watched his chance, and one day when he saw the crane fast asleep, he crept quietly up to him holding a fish-bone. This he stuck right across the root of the crane's tongue.

Then he went off as quietly as he had come; careful, for once, to make no noise.

The crane woke up at last, and when he opened his mouth to yawn he felt like choking. He tried to get the obstruction out of his throat. In the effort he made a queer scraping noise, which was all he could give utterance to. The bone stuck fast.

And to this day the only noise a crane can make is, "gah-rah-gah, gah-rah-gah!" This noise gives the name by which he is known to the blacks.

Folklore and story often anchor our lived experience to a mythic reality. This is especially true for Australian Aboriginal peoples whose stories construct the fabric of their lives. Social laws, morals, and relationships are defined through storytelling, which is also danced, painted, and sung. And Crow is ever-present. He is a trickster character, cunning and intelligent as well as entertaining and naughty. Here, he holds a grudge for having been made black and is pretty sneaky in getting his revenge. Among some tribal groups, Crow represents half of two social moieties, the other half being the Eaglehawk moiety. For this reason, group members personally connect with him. He is much more than a bird. He represents history, ancestral ties, and a connection to place. It is said among the tribal peoples of New South Wales that,

> Towards the end of his stay on Earth, he [Crow] was travelling down the Murray river when he came across Swamp Hawk. Crow decided to play a trick on the bird. He planted echidna quills in the deserted nest of a kangaroo rat and got Swamp Hawk to jump on them. One of the interesting things about many of Crow's tricks is that they benefit the person he plays them on, and in this case Swamp Hawk was pleased, for the quills grew into his feet and he found that he could catch kangaroo rats easily.
>
> Crow continued on his journey and became caught in a storm. The rain lashed down and he felt cleansed by it. It was then that a voice was heard. It was Biame the All-Father. He took the old Crow up into the sky where he became the star Canopus.[8]

When they look to the night sky, the Aboriginal peoples see Crow still with them in the celestial stars.

This story was collected by Catherine (Katie) Langloh Parker (1856–1940) in 1897 but a version of it is known and valued today as a traditional Nhunggal (northwestern New South Wales) story. Parker grew up around the Euahlayi people. She took an interest in their lives and stories, and published two books of myths. She took great pains to record the stories accurately as told, going back and forth between teller and

***The Eaglehawk & The Crow*. 1996. Rover (Julama) Thomas, Kukaja-Wangkajunga. Australia. Natural earth pigments on Belgian linen, 23.6 × 35.4 inches. The artist explains that Eaglehawk (*koondooroo*) is watching Crow (*wonjanali*) to ensure that he doesn't get to the heart of dead Kangaroo (*malu malu*) as this is eaglehawk's favorite food (*mungari*). In Australian aboriginal mythology, Crow is always trying to outwit his many competitors (© Rover Thomas / Copyright Agency. Licensed by Artists Rights Society [ARS], New York, 2023. Image courtesy of Japingka Gallery, Fremantle, Western Australia).**

translator until everyone was satisfied. Her work is valued for what it preserves of the past but her attitude and the commentaries in her work show all the signs of superior colonial sentiment. She was a woman yoked to her time.

## Why Crow Is Black and Men Find Precious Stones in the Earth

Told among the Palung, Burma. Retold by Natalia Belting.

Once there was a Naga princess named I-ran-ti who grew tired of living under the earth in the country of the Nagas, the country of the serpent-dragons.

"I shall go and live on the earth," I-ran-ti said. She looked at herself in a great diamond. "I shall be the most beautiful woman on earth," I-ran-ti said, and at once she was no longer a serpent-dragon but a woman.

Her skin was the color of the brown earth when the setting sun shines on it; her hair was as black as the shadows under the trees at noon-day. Her gold hairpin was like the sun itself in its brilliance. Her garments were silk, thin as tissue, embroidered in crimson and purple stripes. Her black velvet cap had ornaments of silver wire, and over the cap she wore a silk scarf that was stiffened so that it looked like the hood of the Naga serpent-dragons, with jewels in place of the serpent scales.

I-ran-ti was pleased. "I wish to be on earth," she said, and at once she was walking by the side of a quiet stream in the middle of a far-stretching plain, and the scent of the grasses and the flowers was sweet.

The Sun saw I-ran-ti. He drove his chariot close. "She is the most beautiful woman in the world," he said. He brought his chariot down to the earth and drove along beside her as she walked.

I-ran-ti looked at him. She saw a man, tall and handsome. She saw a man in gold and silk who stopped his chariot and got out and walked with her.

"There is no woman so beautiful," the Sun said to I-ran-ti.

"There is no man so handsome," I-ran-ti answered.

They walked together. The Sun took I-ran-ti's hand, and I-ran-ti sang in her happiness. The Sun was filled with love as all men are filled with love when they walk with a beautiful woman in the warm spring air.

"Thou art my true love, my beautiful I-ran-ti," the Sun whispered.

"Thou art my true love," I-ran-ti answered.

But the time came when the Sun said to I-ran-ti, "My beloved, I must leave you. I must go back to the sky. If I do not, the crops will not grow, and men will have no food. The earth will be cold. It will die. I will come back. I will leave my chariot with you as a pledge of my return, as a pledge I will return before our children are born."

I-ran-ti walked by herself, and every day looked upon her lover as he crossed the sky, and sang in her happiness, and sometimes in her sadness because the Sun did not return to her. Then her children were born. They were not like the children of mortals, for I-ran-ti was a Naga serpent-dragon. The children were eggs. She laid them carefully in the chariot, and watched them, and waited for the Sun to come.

The Sun did not come. He looked down on I-ran-ti where she waited for him, in the chariot, caring for the eggs, and he said, "She has deceived me. She is not a woman. She is a Naga, a serpent-dragon." The Sun excused himself for his fickleness.

"I should give her a gift, however," the Sun said. He called Crow. Crow came, and his gold feathers were even more gold when the Sun looked at him. "Take this little bag," the Sun ordered. "Take it to I-ran-ti. There is a message inside, and a gift."

Crow flew down to earth. The way was long. Crow grew tired. He came upon a flock of birds eating. "I am hungry," he said. He took the little bag off from around his neck and hung it on a branch and joined the other birds.

While he ate, a man came by and saw the bag where Crow had hung it. He took it down and opened it. "Good fortune has visited me," he said when he saw the jewel that was in the bag. He looked around. "I do not know who left this here. It is clear he has forgotten it." The man picked up a pebble and put it in the bag and hung it back on the branch. The man went away.

When he finished eating, Crow flew off with the bag to I-ran-ti. "My master, the Sun, sent this gift," Crow said.

I-ran-ti was filled with joy. She opened the bag. She read the Sun's message, "I shall not return, but I give you a magic gem so that you will remember the happiness we shared."

I-ran-ti was filled with sudden grief. She took the pebble out of the bag. I-ran-ti was filled with rage, with all the rage of a Naga serpent princess. She hurled the pebble at Crow. She stamped upon the bag. She tore the message to shreds. "This is no magic gem," she said. "The Sun mocks me."

She took the eggs out of the chariot and threw them. One egg fell upon the rock hills and broke, and whatever was touched by its contents became jade, and the mountains were filled with jade. The second egg fell upon the mountains, and whatever was touched by its contents became rubies and emeralds and sapphires, and men mine the mountains for them.

I-ran-ti tore the silk scarf from her head in her anger. Her gold hairpin fell upon a high mountain and pierced it, and men mine gold and wash gold dust out of the mountain streams.

I-ran-ti became a Naga serpent once more. She shed her human form. She went back to the country of the Nagas, deep in the earth.

As for Crow, the Sun in his anger painted Crow black. "It was your fault the magic gem was lost," Sun said. "From this day you will live on earth. You will be the ugliest bird on earth. You will crow, not sing. Other birds will be afraid of you. Men will dislike you."

The mythological Crow often makes mistakes. He bungles assignments, acts rashly, is habitually lazy (and sometimes negligent), but he is usually good-hearted. This story associates the character with themes that will come up in later chapters—the sky, the sun, the messenger—but it is fundamentally an origin story. We learn how an entire landscape became rich in precious stones (Burma harbors a wealth of gems: jade, ruby, sapphire, diamond, tourmaline, peridot, topaz, garnet) and we discover how Crow turned from a shiny golden hue to dusky black. Origin stories exist throughout the world as explanations for the way things are. Today we believe in scientific explanations but in the absence of scientific facts we look to the supernatural forces that shape our world. Instead of plate tectonics, we have the sun and the serpent acting upon the earth. In many ways we connect with Crow more than with the resplendent gods in the story. It could easily have been ourselves who sat down with friends to dreamily eat lunch.

The Palung are an ethnic minority found in the northern parts of the Shan state in Burma as well as in Yunnan Province, China and Northern Thailand. They make their living through agriculture, fishing, and logging. The Palung practice Buddhism, but they have a rich tradition of storytelling, music, and dance that blends their religion with ancient traditional animism.

## The Crow and the Soap

Persia. Collected and translated by Alice G. Kelsey.

One day Fatima, the wife of Mullah Nasr-ed-Din, went to wash her clothes in the ditch that brought water to the village. She carried a flat stick to pound them clean. Because she was very lucky that day, she carried also a small piece of yellow soap to rub on them. This soap was very precious. It was expensive if she bought it in the market, and it was hard to make if she saved fat to boil in the big kettle in her yard. But what a difference it made as she pounded with her flat stick to get the clothes clean in the cool flowing water of the ditch.

Usually there were other women washing their clothes. Today Fatima happened to be alone except for a big black crow that perched in the branches of a willow tree and greeted her with an angry "Caw, caw, caw!" Fatima would have liked to have women to gossip with her. On the other hand, had they been there they would have asked to share her precious soap. The Mullah had brought it to her as a special surprise just yesterday when he had sold his figs for a good price. She must make it last as long as possible.

"Caw, caw, caw!" scolded the big black crow.

Slap—slap—slap went Fatima's stick as she pounded her clothes in the flowing water. She paid no attention to the crow in the willow tree.

Suddenly the big black bird swooped toward her. Fatima threw up her hands to shield her face. But it was not the woman that interested the crow. With his sharp beak he picked up the piece of soap. He flapped his ragged black wings and flew to the top-most branch of the willow tree.

"My soap!" screamed Fatima. And a very good voice she had for screaming. "My soap! My soap! Fly back here with my soap!"

Fatima's voice was so shrill and so loud that it was heard easily in the village. The Mullah, talking with his friends Jafar and Mehmet Ali near the gate of the village walls, recognized it. His good wife must be in great danger! He went running to rescue her, his long coat fluttering behind him as he ran.

"What is the matter?" he called. "I will save you, my Fatima. What has happened?"

"My soap! My soap!" Fatima's scream was near enough to him now that he could understand her words.

"What about your soap?" He stood beside her, wondering how a mere piece of yellow soap could have given her such terrible pain.

"It is gone," sobbed Fatima. "A big crow grabbed it in his horrid beak and flew away with it. Ah wahi! Ah wahi! It was the very best soap. It was the very piece of soap you gave me yesterday. When can I ever have another?"

At first the Mullah felt almost as badly as Fatima did about the stolen soap. He knew how much he had paid for it in the bazaar only the day before in spite of his best bargaining. Gazing upward, he could see the black of the crow's feathers in the topmost branches of the willow tree. Suddenly he realized that everything had happened for the best after all.

"There, there!" the Mullah soothed his screaming wife. "Did you notice the color of the crow's coat?"

"Black of course!" Fatima stopped crying to wonder why her husband should ask such a simple question.

"Yes, blacker than the dirtiest of our clothes," said Mullah Nasr-ed-Din. "He needs the soap much more than we do. Let him keep it!"

This amusing story equates the blackness of the crow with uncleanliness. The corvid habit of eating just about anything alive or dead certainly brings this to mind. But scavenging can be thought of as cleansing just as easily as the opposite—and black crows are not the only bird species to pick at bones. Nevertheless, there is something about a crow's blackness that is unsettling ... as well as beautiful. Persian poetry sometimes uses the raven's black feathers in similes to convey the alluring darkness of a woman's hair or the special quality of a starless night.[9]

The Islamic religion is dominant in the area once known as Persia (historically a large part of southwestern Asia, now associated with Iran). It regulates all aspects of daily life, including diet. Qur'anic law designates birds with a biological crop as clean and those without a crop as unclean, those with webbed feet as clean and those that soar or glide as unclean.[10] Grain eaters receive no hospitality, but crows are said to have medicinal value. The meat of a black crow is recommended for malaria or prolonged illness, and it may be its very blackness that gives it medicinal power.[11] The crow and raven's magical powers are addressed in Chapter 7.

In nearby northern Pakistan, we discover a fabled white crow among the Kalasha people. During their winter solstice festival called *Chaumos*, which blends Islamic and traditional beliefs, the white crow brings wealth and fertility. Near the end of the festival, young Kalasha girls go to a temple in the darkness of night. They decorate the ritual space and sing a prayer addressed to the white crow:

> Make us Prolific with offering, come O little Crow.
> Show us Happiness upon Happiness, come O little Crow.
> Give us wellbeing, come O little Crow
> Bring sacrifices of the He-Goats, come O little Crow....
> Bring wealth, come O little Crow.
> Pray to 'Allah Pak' for us, come O little Crow.[12]

At the end of the prayer the girls throw cracked walnuts down the slopes near the temple for actual crows to consume.

Black versus white.

## Raven and the Fear of Growing White

Duane Niatum, Jamestown S'Klallam, Washington State.

> When the legends cannot feed the village fire,
> When mother spruce answers no child in the dark,
> When hawk fails to reach his shadow on the river,
> When First Woman beats hummingbird to the earth,
> And salmon swims the river until his bones shatter,
> When otter steals the long-awaited promises of stars,
> And blue jay stops naming each new storm,
> He will end his fear of growing white.

Duane Niatum (b. 1938) is a Native American poet, author, and playwright. Although he studied with such pillars of American poetry as Theodore Roethke and Elizabeth Bishop, Niatum's work is animated by the landscape of the Pacific Northwest and the stories of his people. His poetry references things that hold deep symbolic significance and long histories among the Klallam. In this poem, he uses black and white as metaphors to address the deracination he sees happening around him. Raven (i.e., Niatum's people) does not want to be "white," but the poem drives in that direction. Niatum structures the work almost like a chant, using "When" and "And" repeatedly. It ends in an uneasy conclusion. When (or if) these events happen, Raven's fear of growing white will cease—because he will have already become white.

***White Raven*. 2023. Iris ten Brink. Digital image. A haunting image presents an impossible scene. The viewer is left to wonder if they are in a dream or having a hallucination (© Iris ten Brink).**

## *A Bird Artist: Edward Lear*

Looking at a crow or raven from an ornithologically artistic point of view opens an entirely new window onto its world. The art of bird illustration is vast in scope and goes back as far as the thirteenth century in Europe when the goal of capturing the demeanor and details of birds appealed to all sorts of newly minted ornithologists. Bird artists routinely included the crow and raven in their portfolios. John James Audubon (1735–1851) is perhaps the most well-known of bird illustrators and he wrote of the raven: "There, through the clear and rarefied atmosphere, the Raven spreads his glossy wings and tail, and, as he onward sails, rises higher and higher each bold sweep that he makes, as if conscious that the nearer he approaches the sun, the more splendent will become the tints of his plumage." Adding, "On the ground the Raven walks in a stately manner, its motions exhibiting a kind of thoughtful consideration, almost amounting to gravity."[13]

In addition to Audubon, Edward Lear (1812–1888) was an illustrator of his time. In his *Raven, Corvus Corax*, we see the bird with upturned head in a three-quarter view—as if alert perhaps to a new sound or startling movement. The beak is firmly shut but there is a gleam in the wide, focused eye. Although the inky pelt is striking against the plain background, we notice remarkably minute details in the feathers as well as a pattern that makes its way down the side of the body. Strong, almost frightening, talons secure this bird in its place.

Lear grew up in an age of intense interest in botany and ornithology. A flurry of activity followed the publication of works by Carl Linnaeus (1701–1778) on plant and animal classification and the voyages of discovery by those like Captain Cook in the 1760s. Scientists, both the professional and the amateur, rushed to collect, identify, label, and describe all types of new species.

This was a good thing for Edward Lear who, as the twentieth child of a stockbroker gone bankrupt, was forced to earn a living from the age of 15 through his drawings. He honed his skills to the point where he became an expert. He was one of the first artists to draw living birds (Audubon used stuffed skins) and his talent extended to capturing their unique characters as well as their precise features. Lear was in much demand as a zoological illustrator, contributing to major publications on natural history and making a name for himself as a lithographer. The wealthy and renowned ornithologist John Gould noticed his work and hired him to assist in the publication of *The Birds of Europe*, an enormous five-volume tome with 448 plates. Scholar Susan Hyman writes: "they are certainly among the most remarkable bird drawings ever made … it is evident that Lear endowed them with some measure of his own whimsy and intelligence, his energetic curiosity, his self-conscious clumsiness and

***Raven, Corvus Corax*. 1837. Edward Lear. England. Lithograph with hand-finished color. In John Gould, *The Birds of Europe*, Volume III. Edward Lear is known by young and old for his light-hearted limericks and whimsically illustrated poems (i.e., "The Owl and the Pussycat" and "The Jumblies"), but his early career was spent as an accomplished, self-taught, naturalist and artist (General Research Division, The New York Public Library).**

his unselfconscious charm."[14] But Gould was reluctant to share credit or accolades and barely mentioned Lear's contributions.

Lear suffered from asthma, epilepsy, depression and poor eyesight. At age 25, he disappeared from the ornithological world and made himself famous for authoring books of limericks and comical verse. He died at the age of 75.

**Photo portrait of English artist and writer Edward Lear. 1862. While he did not embark on scientific expeditions, Lear spent a great deal of time outdoors and at zoos studying all types of animals. He found that his skills were especially suited to depicting birds (Ian-Dagnall Computing/Alamy Stock Photo).**

Edward Lear embraced the blackness of the raven; most myths and stories from around the world tend to dilute the enormous power inherent in its dark shape. A fatal flaw, in personality or strategy, seems to have caused its coat to become black. The tragedy (and comedy) of being black, however, becomes the spark of creation, the incitement of beginnings, and the explanation for how things are in this world. This event starts things off.

In the Western world, we associate the color black with evil, impurity, or sin. Saint Gregory is supposed to have said: "The raven is every learned preacher who proclaims loudly while he wears the memory of his sins as if they were a kind of blackness of color."[15] But the stories that follow show us a black bird that is powerful, magical, and in every way divine.

# 2

# A Creator

"There is wisdom in a raven's head." So goes a well-known Scottish Gaelic proverb, and the Scots may be on to something. Corvids are incredibly intelligent birds. Scientists have been studying their aptitudes for decades, knowing that these feathered creatures have a significantly larger brain size than would be expected for their body dimensions and exceptionally large forebrains, an area involved with sensory processing, thought analysis, and behavioral flexibility. But does this necessarily translate into intelligence, and to what extent? Are my backyard crows smart? Could they ever create something out of nothing? Maybe, maybe not.

## *Field Notes: Tool Use*

One sign of cognitive acumen is the use of tools. The ability to make and manipulate a tool was once thought to be the hallmark of human society alone but times have changed. Corvids have been documented doing both, using and making tools. Ornithologist Lawrence Kilham, for example, watched his pet raven Raveny sweep aside a pile of sand using a flat piece of wood as his tool.[1] In his laboratory at the University of Chicago, Dr. Benjamin Beck witnessed a hungry crow pick up a small plastic cup, dip it into a water trough, carry it across the room, and empty the water onto dried out mash to moisten it for consumption. This astonishing feat had never been taught to the bird.[2]

NYU graduate student Joshua Klein designed a vending machine at the Binghamton Zoo in New York that would put crow intelligence to the test. He offered a flock of crows peanuts and coins, in a dish that was attached to the machine. He then took the peanuts away. As they searched for the missing peanuts, the birds pushed the coins out of the dish into a slot, causing more peanuts to be released into the dish. The Binghamton crows quickly learned to drop coins into the slot to get the peanuts and within a month the crows were scouring the ground for loose change.[3]

Experiments have proved that corvids are born with tool-using abilities, but tool making is another story. The New Caledonian crow, a native of New Caledonia and the Loyalty Islands in the Pacific Ocean, is the poster-bird for tool-making skills. These birds are known to create new things—to shape leaves and twigs into tools that they then utilize to obtain different kinds of food. Biologist Gavin R. Hunt has watched the crows make two specific kinds of tools. For a "hook" tool, they employ their wide beaks to carefully pull a twig away from a branch, then smooth its

**A New Caledonian crow uses a tool to retrieve food. The New Caledonian crow has an advantage over other crow species—it has a straighter bill and bigger, more forward pointing eyes that give it better binocular vision (Auscape International Pty Ltd / Alamy Stock Photo).**

bark, and finally nip it in such a way as to create a hook at the end. The hooked end is thrust into tree holes or dead wood to extract grubs. Another kind of tool is made from the stiff leaf of a spiny pandanus plant. The birds hold the leaf while it is still attached to the plant and take successively deeper bites from it so that it is properly "stepped" to a fine point. They then bite off the finished product. The pointed end is inserted into holes and the spiny serrations on the plant snag prey as it is pulled up. The crows value their tools; they keep track of them and return from forays to retrieve them. When crows change foraging sites, they bring their tools with them.[4] Mr. Hunt observed different communities of crows make slightly different kinds of pandanus tools. It appears to be a learned, "cultural" tradition. Hand-raised baby crows never managed to make any tools.[5]

New Caledonian crows can even combine tools when necessary. In 2007, graduate student Alex Taylor and his colleagues at the University of Auckland, New Zealand documented New Caledonian crows using a short stick to retrieve a longer stick that could then be manipulated to retrieve food. The sequence of actions had not been taught to the birds but they figured it out. One crow looked at the setup and performed the task within 2 minutes. The ability to do this surprised researchers because it involved a high level of both foresight and reasoning.[6]

Dr. Bernd Heinrich has spent his career studying ravens, documenting their abilities to imagine solution possibilities and apply logic to problems. He devised a test in which he brought out a length of hollow pipe and put it in front of a group of Maine ravens. They watched him place a frog in one end of the pipe. The birds quickly jumped down to look at that end—and then ran around to wait at the other end. With this experiment, Heinrich was able to document a process of thinking. His ravens could foresee that the frog would move and that it would probably end up at the other side of the pipe.[7] Heinrich set up another test that involved multiple complicated steps in order to reach a food treat. His ravens examined the situation and then executed the procedure necessary to retrieve the food in as little as 30 seconds on their first try.[8] Dr. Heinrich realized that the birds had absorbed the situation, contemplated the possibilities, and then acted on the best one.

Dr. Heinrich ended his seminal book *Mind of the Raven* with the following statement: "...the bird's world is inordinately more complex than had been previously assumed, ... I conclude that ravens are able to manipulate mental images for solving problems. They are aware of some aspects of their private reality, seeing with their minds at least some of what they have seen with their eyes."[9] Dr. Nathan Emery and Nicola Clayton, of the University of Cambridge, UK, add: "... these birds display similar intelligent behavior as the great apes. [They both] appear to use the same cognitive tool kit: causal reasoning, flexibility, imagination, and prospection."[10]

Even non-scientists who are observant can tell crows are smart. Aesop knew it more than two thousand years ago when he told the story of "The Crow and The Pitcher." My little backyard crows may not be able to create a new tool, but I have seen how intelligently they handle a carcass—pulling long threads of meat to hang delicately along a tree branch. Such a smart bird might also be a powerful bird, potent enough to be a creator, or The Creator. Indeed, the number of creation stories involving Crow and Raven are endless. They appear all over the world and throughout time. Crow and Raven have been known to bring forth the world, fashion the first humans, release the light, and control the tides. What follows is just some of their handiwork.

## The Beginning

Eskimo (Inuit), North Alaska. Collected and translated by Robert F. Spencer.

Men say that the world was made by *tulunixiraq*, the Raven. He is a man and he has a raven's bill on his head. The ground came up from the water. It was brought up by tulunixiraq who appeared on the water. He speared down into the water and brought up the land and fixed it into place.

The first land was a plot of ground, just a little bigger than a house. There was a family that had a house there. There was a man and his wife and their little son. This was tulunixiraq. One day he saw a round thing, a kind of bladder, hanging up over his parent's bed. He wanted to play with it and he asked his father for it again and again. Each time his father said: "No, you can't have it." But the boy begged so much that finally his father gave in. While he was playing with that round thing, he accidentally broke it. Now up to that time, it had always been dark. But now it began to grow light. The father said: "We had better have it night, too, not just daylight all the time." So he grabbed the round bladder before the little boy could damage it further. And that is how day and night began.

Now the father had a kayak. He said: "There is other land far away; how can we reach it?" Then tulunixiraq asked his father to let him go to that other land. But his father said: "No, you had better not go; there is danger there." But at last he agreed to let the boy go. And tulunixiraq paddled a long time over the sea. Finally, his kayak came to a place in the sea where some land was bobbing up and down on the surface of the water. First it would rise up above the surface and then it would sink again. The raven boy was afraid and slowed up. When he came close, the land sank down. As he watched, the land came up again. Now he carried his spear along with him in the kayak. When the land came up again, he speared it and held fast to the lead rope. This fixed the land firmly and it stopped bobbing up and down. Then tulunixiraq got out of his kayak and walked around on the surface or the land. The place where this was done is called Umiat, the landing place (Colville River, 69°30'N., 152°16'W.) even to this day. Because there is high ground there, the raven boy was able to walk about.

After the land became fixed in place, the sea began to move away and there was dry ground all around. And so it is because of tulunixiraq that people are able to live in the world.

Part of a lengthy cycle of Raven stories that explained the nature of everything, this account of the creation of solid ground must have captivated listeners. According to ethnographer Robert Spencer, the telling of tales was a prized skill among the Eskimo (today identified as Inuit) of the North Alaskan slope near Barrow.

> The idea was "to get the story right." If a raconteur deviated as much as by a word in his recital, his skill was considered dubious indeed.... The sense of personal integrity is so strong that unless a man feels he can be accurate, he prefers to keep silent. In telling a story, the speaker could hold his audience spellbound by his skill. He acted out the parts of the characters and his mimicry was applauded as much as the tale itself. Tales were told in winter.... In winter the people assembled in a house or in the karigi (ceremonial house) and several men told tales. There was often a notion of a contest in storytelling....[11]

At the time this myth was collected (1950s), the Indigenous people living in the Barrow area were straddling modernity and traditional ways. They lived in permanent coastal villages and devoted themselves to sea mammal hunting, maintaining complex social and religious practices around the whale. While their traditions were threatened, they preserved the close cooperative ties of kinship that anchor society.

Anyone who called another by a term of kinship was responsible for that individual's welfare and actions. They strengthened their connections and transferred their cultural heritage at storytelling events.

Robert F. Spencer (1917–1992) was an American anthropologist with wide-ranging interests. He spent time with the Keresan Pueblo people of New Mexico (1938–1940), interviewed imprisoned Japanese Americans in what he described as a dire and abjectly miserable relocation camp of Gila River, Arizona (1942–1943), lived among the North Alaskan peoples, studied the cultures of Turkey (1954), resided for a time in a Burmese monastery (1954–1955), examined the poetry of Pakistan (1965–1966), and studied the lives of factory girls in South Korea (1988). His *The North Alaskan Eskimo: A Study in Ecology and Society*, was his most influential work and regarded as being years ahead of its time. It set a new standard in Arctic ethnography and has been used as a basic reference ever since.

**Bird mask. n.d. Alaskan Inuit. Carved and painted wood, 15 × 10 × 8.25 inches. Transformation masks created along the Northwest Coast and Alaska were worn by dancers and performers in ritual ceremonies. At a dramatic moment in the performance, the mask would be suddenly opened to reveal the hidden being underneath—an ancestor, a god, or a spirit being. The performer thus demonstrated the porous boundary between the natural and supernatural realms (Copyright © Phoebe A. Hearst Museum of Anthropology and the Regents of the University of California).**

## How the Earth Was Made and How Wood-Chips Became Walrus

Told by A'ttin'qeu, Maritime Chukchee (An'kalyn), to Waldemar Bogoras at Mariinsky Post, Siberia, October 1900.

Raven and his wife live together,—the first one, not created by any one, Raven, the one self-created. The ground upon which they live is quite small, corresponding only to their wants, sufficient for their place of abode. Moreover, there are no people on it, nor is there any other living creature, nothing at all. No reindeer, no walrus, no whale, no seal, no fish, not a single living being.

The woman says, "Ku'urkil."*

"What?"

"But we shall feel dull, being quite alone. This is an unpleasant sort of life. Better go and try to create the earth!"

"I cannot, truly!"

"Indeed, you can!"

"I assure you, I cannot!"

"Oh, well! Since you cannot create the earth, then I, at least, shall try to create a 'spleen-companion.'"

"Well, we shall see!" said Raven.

"I will go to sleep," said his wife.

"I shall not sleep," said Ku'urkil. "I shall keep watch over you. I shall look and see how you are going to be."

"All right!"

She lay down and was asleep. Ku'urkil is not asleep. He keeps watch, and looks on. Nothing! She is as before. His wife, of course, had the body of a raven, just like himself. He looked from the other side: the same as before. He looked from the front, and there her feet had ten human fingers, moving slowly.

"Oh, my!"

He stretches out his own feet, the same raven's talons.

"Oh," says he, "I cannot change my body!"

Then he looks on again, and his wife's body is already white—and without feathers, like ours.

"Oh, my!"

He tries to change his own body, but how can he do so? Although he chafes it, and pulls at the feathers, how can he do such a thing? The same raven's body and raven's feathers! Again, he looks at his wife. Her abdomen has enlarged. In her sleep she creates without any effort. He is frightened, and turns his face away. He is afraid to look any more.

He says, "Let me remain thus, not looking on!"

After a little while he wants to look again, and cannot abstain any longer. Then he looked again, and, lo! there are already three of them. His wife was delivered in a moment. She brought forth male twins. Then only did she awake from her sleep. All three have bodies like ours, only Raven has the same raven's body. The children laugh at Raven, and ask the mother, "Mamma, what is that?"

"It is the father."

"Oh, the father! Indeed! Ha, ha, ha!"

They come nearer, push him with their feet. He flies off, crying,

---

* Ku'urkil is the name of Raven

"Qa, qa!" They laugh again. "What is that?"

"The father."

"Ha, ha, ha! the father!" They laugh all the time.

The mother says, "O children! you are still foolish. You must speak only when you are asked to. It is better for us, the full-grown ones, to speak here. You must laugh only when you are permitted to. You have to listen and obey."

They obeyed and stopped laughing.

Raven said, "There, you have created men! Now I shall go and try to create the earth. If I do not come back, you may say, 'He has been drowned in the water, let him stay there!' I am going to make an attempt."

He flew away. First, he visited all the benevolent Beings (*va'lrglt*), and asked them for advice, but nobody gave it. He asked the Dawn, no advice. He asked Sunset, Evening, Mid-day, Zenith—no answer and no advice. At last he came to the place where sky and ground come together. There, in a hollow, where the sky and the ground join, he saw a tent. It seemed full of men. They were making a great noise. He peeped in through a hole burnt by a spark, and saw a large number of naked backs. He jumped away, frightened, ran aside, and stood there trembling. In his fear he forgot all his pride in his recent intentions.

One naked one goes out. "Oh! it seemed that we heard someone passing by, but where is he?!"

"No, it is I," came an answer from one side.

"Oh, how wonderful! Who are you?"

"Indeed, I am going to become a creator. I am Ku'urkil, the self-created one."

"Oh, is that so?"

"And who are you?"

"We have been created from the dust resulting from the friction of the sky meeting the ground. We are going to multiply and to become the first seed of all the peoples upon the earth. But there is no earth. Could not somebody create the earth for us?"

"Oh, I will try!"

Raven and the man who spoke flew off together. Raven flies and defecates. Every piece of excrement falls upon water, grows quickly, and becomes land. Every piece of excrement becomes land—the continent and islands, plenty of land.

"Well," says Raven, "Look on, and say, is this not enough?"

"Not yet," answers his companion. "Still not sufficient. Also, there is no fresh water; and the land is too even. Mountains there are none."

"Oh," says Raven, "shall I try again?"

He began to pass water. Where one drop falls, it becomes a lake; where a jet falls, it becomes a river. After that came mountains, smaller pieces became hills. The whole earth became as it is now.

Then he asks, "Well, how is it now?"

The other one looked. "It seems still not enough. Perhaps it would have been sufficient if there had not been so much water. Now some day the water shall increase and submerge the whole land, even the mountain-tops will not be visible."

Oh, Raven, the good fellow, flew farther on. He strains himself to the utmost, creates ground, exhausts himself, and creates water for the rivers and lakes.

"Well, now, look down! Is this not enough?"

"Perhaps it is enough. If a flood comes, at least the mountain-tops will remain above water. Yes, it is enough! Still, what shall we feed upon?"

Oh, Raven, the good fellow, flew off, found some trees, many of them, of various kinds—birch, pine, poplar, aspen, willow, stone-pine, oak. He took his hatchet and began to chop. He threw the chips into the water, and they were carried off by the water to the sea. When he hewed pine, and threw the chips into the water, they became mere walrus; when he hewed oak, the chips became seals. From the stone-pine the chips became polar bears; from small creeping black birch, however, the chips became large whales. Then also the chips from all the other trees became fish, crabs, worms, every kind of beings living in the sea; then, moreover, wild reindeer, foxes, bears, and all the game of the land. He created them all, and then he said, "Now you have food! hm!"

His children, moreover, became men, and they separated and went in various directions. They made houses, hunted game, procured plenty of food, became people. Nevertheless, they were all males only. Women there were none, and the people could not multiply.

Raven began to think, "What is to be done?"

A small Spider-Woman (*Ku'rgu-ñe'ut*) is descending from above on a very slender thread.

"Who are you?"

"I am a Spider-Woman!"

"Oh, for what are you coming here?"

"Well, I thought, 'How will the people live, being only males, without females?' Therefore I am coming here."

"But you are too small."

"That is nothing. Look here!" Her abdomen enlarged, she became pregnant, and then gave birth to four daughters. They grew quite fast and became women. "Now, you shall see!"

A man came, that one who was flying around with Raven. He saw them, and said, "What beings are these, so like myself and at the same time quite different? Oh, I should like to have one of them for a companion! We have separated, and live singly. This is uncomfortable. I am dull, being alone. I want to take one of these for a companion."

"But perhaps it will starve!"

"Why should it starve? I have plenty of food. We are hunters, all of us. No, I will have it fed abundantly. It shall not know hunger at all."

He took away one woman. The next day Raven went to visit them, made a hole in the tent-cover, and peeped through.

"Oh," says he, "they are sleeping separately in opposite corners of the sleeping room! Oh, that is bad! How can they multiply?"

He called softly, "Halloo!, Halloo!"

The man awoke and answered him. "Come out here! I shall enter."

He entered. The woman lay quite naked. He drew nearer. He inhaled the odor of her arm. His sharp beak pricked her.

"Oh, oh, oh!"

"Be silent! We shall be heard."

He pushed her legs apart and copulated with her. Then he repeated it again. The other one was standing outside.

He felt cold, and said, "It seems to me that you are mocking me."

"Now, come in! You shall know it too. This is the way for you to multiply."

The other one entered. The woman said, "It is a good thing. I should like to repeat it once more."

The man answered, "I do not know how."

"Oh, draw nearer!"

He says, "Oh, wonderful!"

"Do this way, and thus and thus." They copulated.

Therefore girls understand earlier than boys how to copulate. In this manner human kind multiplied.

The Chukchee people, from whom this story comes, lived in a sparsely settled area, accessible only by dog sled, boat, or horseback. Mariinski Post, where the story was told, was a small Russian administrative and trade station at the mouth of the Anadyr River, just across from Alaska on the Chukchee Peninsula, Siberia. The name Chukchee means "rich in reindeer" but the Maritime Chukchee called themselves *añqa'lit,* or "sea people." Like all Indigenous peoples of the area, they were supremely adapted to the harsh Arctic environment. They lived in semi-subterranean houses and traveled by skin-covered, wood-framed boats to hunt sea mammals. They were also wonderful storytellers, drawing upon multiple versions of epic stories about how Raven created the world. This one is particularly amusing in its account of the ways in which land, water, mountains, animals, and humans were created. Everyone in the story is portrayed as needing to figure things out at the beginning of time. Raven needed to be shamed by his wife and then encouraged by men to create the features of the earth (which came directly from his body's waste). The men, in turn, needed to learn how to procreate. The women, on the other hand, were the capable ones. They brought forth life as we know it today.

Known in the West as Waldemar Bogoras, Vladimir Germanovich Bogoraz (1865–1936) was a monumental figure—a Russian revolutionary (dismissed from Law School for his activities) as well as a poet and citizen anthropologist. He was imprisoned for distributing anti-government propaganda and then exiled

**Chukchi ceremony, four birds. 1901. Waldemar Bogoras. Photograph. Siberia. This is one of over 700 photographs Waldemar Bogoras took during his stay among the Chukchee. It documents a small winter ceremony. The constructed ceremonial net appears to be designed to capture birds (American Museum of Natural History).**

to northeastern Siberia from 1889 to 1899 where he began to record the life, language, customs, and beliefs of the remote Chukchee people living there. Bogoras was hired in 1900 to participate in an enormous undertaking sponsored by the American Museum of Natural History in New York, the Jesup North Pacific Expedition, whose goal was to document cultures across both sides of the North Pacific—Siberia, Canada, Alaska, and the American Northwest coast. The project lasted six years; Bogoras spent thirteen months of it traversing the brutal terrain of northeast Siberia. He eventually ended up as a curator at the American Museum of Natural History where he embarked on his seminal work, *The Chukchee*. However, Bogoras found that he couldn't stay away from his motherland for long and returned to Russia. Amid his tumultuous life in Russia (he organized the First Peasant Congress and the Labour Group, got involved in the first Russian Revolution of 1905–1907, and created the first Russian ethnography department at Petrograd University), Bogoras left an impressive legacy in his detailed and thorough work on the Chukchee, a people barely heard of in the West.

## LIGHT COMES TO MANKIND

Told by Ivaluardjuk, Iglulik. Canadian Northwest Territories.
Collected and translated by Knud Rasmussen.

During the first period after the creation of the earth, all was darkness. Among the earliest living beings were the raven and the fox. One day they met, and fell into talk, as follows:

"Let us keep the dark and be without daylight," said the fox.

But the raven answered: "May the light come and daylight alternate with the dark of night."

The raven kept on shrieking: *"qa'Mffc qa' or n_!"*

(Thus the Eskimos interpret the cry of the raven, *qa'° r rl,* roughly as *qa'° q,* which means dawn and light. The raven is thus born calling for light.)

And at the raven's cry, light came, and day began to alternate with night.

The Iglulik people were always small in number (between 400–600 members in the 1820s as well as the 1980s) but they ranged over a wide territory. When Knud Rasmussen encountered them in the first part of the twentieth century, they were already acculturating to Western ways. The Iglulik traditionally engaged in seasonal activities—whale, seal, and walrus hunting in the summer, caribou hunting and fishing in the fall, sea ice hunting in the winter and spring. Today they are employed in the mining and oil industries.

This story was recorded by Knud Rasmussen during his Fifth Thule Expedition (see Chapter 1). In his notes, Rasmussen remarks that the storytellers among the Iglulik seemed to take for granted that all their myths were well known, hence they might start off in the middle of a tale or leave out whole episodes if they felt like it. The actions in the stories were considered to be true, to have occurred, and to be a part of their tribal history. In addition to plants, animals, and humans, their world was filled with supernatural beings, as well as ghosts, spirits, and souls, who all participated in daily life.

Rasmussen made a point to inform his readers that in recording the myths he and his assistant listened to them several times and from different people before committing them to writing. They then read the narratives back to the storytellers to make necessary corrections. This episode is a far cry from the Old Testament "Let there be light" story. Here we have a spirit animal using very animal-like techniques (his "*qa'°q*") to bring on the day.

# The Origin of the Raven and the Macaw, Totems of Winter and Summer

Zuni (A:Shiwi), New Mexico. Collected and translated by Frank Hamilton Cushing.

He who was named Yanáuluha carried ever in his hand a staff which now in the daylight appeared plumed and covered with feathers of beautiful colors—yellow, blue-green, and red, white, black, and varied. Attached to it were shells and other potent contents of the under-world. When the people saw all these things and the beautiful baton, and heard the song-like tinkle of the sacred shells, they stretched forth their hands like little children and cried out, asking many questions.

Yanáuluha, and other priests (*shiwanáteuna*) having been made wise by the teaching of the masters of life (god-beings) with self-magic-knowing (*yam tsépan ánikwanan*), replied: "It is a staff of extension, wherewith to test the hearts and understandings of children." Then he balanced it in his hand and struck with it a hard place and blew upon it. Amid the plumes appeared four round things, seeds of moving beings, mere eggs were they, two blue like the sky or *turkis*; two dun-red like the flesh of the Earth-mother.

Again the people cried out with wonder and ecstasy, and again asked they questions, many.

"These be," said he who was named Yanáuluha, "the seed of living things; both the cherishers and annoyancers, of summer time; choose ye without greed which ye will have for to follow!" For from one twain shall issue beings of beautiful plumage, colored like the verdure and fruitage of summer; and whither they fly and ye follow, shall be everlastingly manifest summer, and without toil, the pain whereof ye ken not, fields full fertile of food shall flourish there. And from the other twain shall issue beings evil, uncolored, black, piebald with white; and whither these two shall fly and ye follow, shall strive winter with summer; fields furnished only by labor such as ye wot not of shall ye find there, and contended for between their offspring and yours shall be the food-fruits thereof.

"The blue! the blue!" cried the people, and those who were most hasty and strongest strove for the blue eggs, leaving the other eggs for those who had waited. "See," said they as they carried them with much gentleness and laid them, as one would the new-born, in soft sand on the sunny side of a cliff, watching them day by day, "precious of color are these; surely then, of precious things they must be the seed!" And "Yea verily!" said they when the eggs cracked and worms issued, presently becoming birds with open eyes and with pin-feathers under their skins,

"Verily we chose with understanding, for see! yellow and blue, red and green are their dresses, even seen through their skins!" So they fed the pair freely of the food that men favor—thus alas! cherishing their appetites for food of all kinds! But when their feathers appeared they were black with white bandings; for ravens were they! And they flew away mocking our fathers and croaking coarse laughs!

And the other eggs held by those who had waited and by their father Yanáuluha, became gorgeous macaws and were wafted by him with a toss of his wand to the far southward summer-land. As father, yet child of the macaw, he chose as the symbol and name of himself and as father of these his more deliberate children—those who had waited—the macaw and the kindred of the macaw, the *Múla-kwe*; whilst those who had chosen the ravens became the Raven-people, or the *Kâ' kâ-kwe*.

Thus first was our nation divided into the People of Winter and the People of Summer. Of the Winter those who chose the raven, who were many and strong; and of the Summer those who cherished the macaw, who were fewer and less lusty, yet of prudent understanding because more deliberate. Hence, Yanáuluha their father, being wise, saw readily the light and ways of the Sun-father, and being made partaker of his breath, thus became among men as the Sun-father is among the little moons of the sky; and speaker to and of the Sun-father himself, keeper and dispenser of precious things and commandments, Pékwi Shíwani Éhkona (and Earliest Priest of the Sun). He and his sisters became also the seed of all priests who pertain to the Midmost clan-line of the priest-fathers of the people themselves "masters of the house of houses." By him also, and his seed, were established and made good the priests-keepers of things.

There is much to ponder in this Zuni telling. As with Australian Aboriginal peoples, the Zuni of New Mexico were divided into two original groups. We learn how it occurred here when the great Yanáuluha gave the people a choice over their future. Raven is central to the important moment, but he seems to have been the less desired; he is described in an unfavorable light. Nevertheless, we discover that the Raven Winter People were many and strong; certainly not the worst fate that can befall a person. The story is a cautionary parable—those who rushed to choose the brightly colored bauble of an egg ended up becoming descendants of the ugly, harsh sounding raven while those who held back and had some measure of self-control found themselves in the beautiful macaw clan. In the end, they balance one another. They are both essential parts of the universe as expressed in the seasons. At the conclusion of the story, we see how everything ties together. The two clans are linked to two animal species and two seasons while Yanáuluha, their father, links himself to the Sun-father as well as to all the Zuni priests who come after.

"The Origin of the Raven and the Macaw" is told in an antiquated style. It sounds almost biblical in tone; as if it is coming from the mouth of a nineteenth century Western person. Sadly, this was often the case for early folklorists who recorded Indigenous myths and stories. They were writing for a Western audience who, they supposed, could not understand or relate to traditional orations. Their renditions almost certainly misrepresented much of the tone and meaning of the original myths. Frank Hamilton Cushing (1857–1900) was, however, a pioneer in the field of immersive anthropology. He was one of the first professional anthropologists to live with the people he was researching, developing long-term relationships with them and participating in every aspect of their lives. Some have argued that Cushing

**Mask of Säl'imobiya (Warrior) of the Nadir: front view. 1901. August Hoen. Chromolithograph. Plate LVII in Matilda C. Stevenson, *The Zuñi Indians*. The Warrior of the Nadir is part of the Council of Gods. His counterpart is the Warrior of the Zenith. Both individuals appear wearing masks with collarettes of raven plumes, painted bodies, anklets and wristlets of spruce, a war pouch, strings of black and white corn, and bunches of giant yucca leaves in each hand. They perform with other directional gods during the night ceremonies of the Council of the Gods (The Reading Room/Alamy Stock Photo).**

forced himself upon the pueblo dwellers and did not always respect the secret aspects of their sacred lives. He took liberties and comported himself in ways that were consistent with his dominant White position. Nevertheless, while his Western privilege certainly affected his perceptions, he did not adopt the typical stance of the time as a detached observer and superior commentator. The depth and breadth of his work helped establish the legitimacy of participant observation as a research strategy and he pioneered an idea called "reflexive anthropology" whereby researcher and subject share their cultures and experiences as equals.

## Why the Tides Ebb and Flow

Tahltan, British Columbia. Collected and translated by James A. Teit.

Now, the people in many parts of the country had no food. Game and all kinds of food were in possession of a few persons (or families), who alone controlled these things. Thus, many people were constantly starving.

Raven followed the shores of the ocean in his canoe. As he went along, he noticed many things underneath the water which the people could eat; but, owing to the depth of the water, this food was out of reach. At last he came to a large man sitting down on the edge of the water.

He asked him why he was sitting there; and the man answered, "If I get up, the ocean will dry up."

It seems, he was sitting on a hole in the earth through which the water poured when he arose. Raven told him to get up, but he would not do so.

Then Raven took him by the hair and pulled him up so far that he was able to put a rock underneath him. The rock was sharp; and when the man sat down again, it hurt him, and he jumped up farther. Raven then put a larger sharp-pointed rock under him. Thus he continued until the man was sitting almost upright. The ocean went down a long way, and exposed the beach.

Raven said to the man, "Henceforth you must get up twice a day, and let the sea go down as far as it is now, so that people may obtain food from the beach. Then you will sit down again to let the water gather and come up. If you promise to do this, I shall not kill you."

At last the man promised, and thus the tides were made. The people were able to find many kinds of food in abundance along the shore, and they no longer starved.

Tahltan Territory lies in the northwestern corner of British Columbia around the upper reaches of the Stikine River and its tributaries. Waterways supported vigorous trade between coastal nations and interior nations. First contact with Europeans occurred in 1838 when Robert Campbell of the Hudson's Bay Company arrived to set up operations in the territory. This was followed by the frenzy of the Yukon gold rush. By the early 1900s, the Tahltan population had been devastated by the smallpox, measles, influenza, and tuberculosis introduced by Europeans. At its lowest point, Tahltans numbered under 300 people; today there are about 4,000 members.

Stories about Big-Raven (*TsE' sketco*) are embedded in the physical landscape. They are carved onto totem poles, sculpted into pipes, and painted on canoes. They are sung under the stars and whispered behind closed doors. Landmarks, large and small, identify Raven's comings and goings. To travel through the land is to travel in his footsteps and to witness the results of his work and adventures. Visit an inlet and you visit the place where Raven quieted the sea so he could land his canoe; stop on a cliff and you stop where Raven told Bear where to find good fish; gaze out at Whale Island and you are looking at *the* whale that Raven harpooned. For this reason, accounts concerning Raven hold deep cultural and historical importance for the people. He is the touchstone of identity. He is the great creator and essential provider.

Big-Raven was said to have been born in Tlingit country, on the coastline to the west of Tahltan country. Some claimed he was of miraculous birth, but the more common story is that he was the youngest of many brothers and began traveling as a transformer when he was quite young. Big-Raven's role was to roam the countryside and live with the different peoples, all of whom had various kinds of powers and knowledge that they protected from others. Raven's mission was to make this knowledge the common property of everyone by obtaining possession himself and then giving it away to others. He made sure that everything of value to mankind did not remain the sole property of any particular family or group.

Raven worked up and down the coast among the Tlingit and then ascended the rivers into the interior Tahltan territory. He went up one river after another and tired himself out. When Raven's work was finished, he traveled out to sea toward the setting sun, and disappeared. No one knows where he went, or where he is now, but some people believe he lives with *Kanu'gu* (an ancient of ancients and a sea deity) and other great gods or chiefs, on an island or country way out in the ocean, where the weather is made.

James A. Teit (1864–1922) was a self-taught anthropologist, photographer, and guide who migrated from Scotland to make his home in British Columbia. He participated in the American Museum of Natural History's ambitious Jesup North Pacific Expedition of 1897–1902, collecting information about First Nations peoples of the Northwest Coast. Teit was fluent in the local language and, having married a local woman, a member of the community. Because he had secured their trust, Teit was able to acquire an unsurpassed amount of information. The success of this portion of the Expedition is attributable to his ethnographic work. Years later, Teit became a tireless advocate for Indigenous peoples' human and territorial rights. After obtaining approval from the chiefs, he vigorously and successfully campaigned on their behalf.

## Doru the Birdman

Efe, Ituri Rainforest, Democratic Republic of Congo.
Collected by Paul Schebesta.

One day, while Tore* was sitting on a swing, an Efe (pygmy) who had lost his way in the forest, appeared right on the spot where Tore's fire was burning. Tore's mother was dozing, so the pygmy stole the fire and fled with it. Awakened by the cold, however, the old woman noticed the theft and called for her son. Tore consoled her, saying he would soon catch the thief, and then swung himself after the pygmy, captured him and took the fire away from him.

The Efe went home and told of his adventure. One of his brothers, who was more daring and clever than he, was then asked to go and steal the fire. He went his way and found the little old woman sitting by the fire, and he, too, succeeded in getting it. But everything happened in the same way as before. He, too, was recaptured by Tore and the fire was taken away from him.

Now there was a powerful Efe by the name of Doru, who had magic power, and he set out to steal the fire in his own way. He clothed himself with the feathers of the sacred bird Tawa, the raven, and began to hop and fly "as high as the heavens and as far as the horizon." At a moment when it was unguarded he leapt upon the fire, stole it and flew away with it.

In answer to the cries of his mother, Tore came, consoled her again, and set his swing in motion. The chase began and continued over mountain and valley, up to the heavens and down to the chasms. Tore swung with all his strength, but in vain.

"Doru," he cried at last, exhausted, "thou art my brother. We are of equal birth and were born of the same mother!" Then he clung to the Hungbu tree, and called in a loud voice for his mother. But when he returned he found her numbed with cold and realized she was dead. Then Tore said in his bitterness:

"For this the people shall be punished by death."

Doru, on the other hand, arrived safely at the camp and shared out the fire, for which he was richly rewarded. Everyone gave him a maid to wife, but, alas, very soon one man died, and then another. The people certainly had the fire, but death had also come to them.

Once again, Raven is involved in the beginning of important things—in this case, seeming opposites—life-giving fire and annihilating death. A shaman-like

* Among the Efe (Pygmies), Tore is alternately called "the Spirit of the forest" and "Man of the forest." He is the ever-present creator and provider of all things. He sees and hears everything in the forest and informs the ancestors.

individual, he makes himself into a raven in order to steal fire but his theft has terrible unintended consequences.

At the time this story was recorded, the Efe lived in the vast tropical rain forest of what was once the Belgian Congo, a place of gigantic water-laden trees, eerily filtered light, and mysterious sounds. Remote and self-sufficient, these hunter-gatherers navigated their world by bringing down animals with bow and arrow, collecting roots, mushrooms or fruit, and harvesting honey and termites. They were known to be a small but joyful people, full of playfulness. On moonlit nights, they socialized together by drinking alcohol, dancing wildly, playing games, and telling stories about the forest spirits and their ancestors.

Paul J. Schebesta (1887–1967) was the first person to provide a rich description of Efe culture. A missionary, linguist, and ethnographer, Schebesta visited the Efe during four expeditions between 1929 and 1955. Although a devout Christian, he was praised for his "functionalist" approach to ethnography. He provided no empirical generalizations in his reports, only situational descriptions. Nevertheless, a Western and Christian background undoubtedly affected Schebesta's work. This is evident in the last remark he makes at the end of the story where he calls Tore "the ever-present creator and provider of all things." Masato Sawada of Kyoto Seika University, who also conducted research on the Efe (1985–1998), states: "I believe Schebesta should have to explain why the 'creator-God' was called 'Father' or 'Grandfather.' It might be natural for Schebesta, a catholic priest, to call 'creator-God' 'Father.' But for me it's not natural at all." Sawada found no evidence of a supreme being or creator-god among the Efe, instead finding many kinds of supernormal beings at work in their world.[12] The Western lens through which almost all early explorers, anthropologists, and missionaries approached new cultures is only now being evaluated.

## Crow Blacker than Ever

Ted Hughes, England.

When God, disgusted with man,
Turned towards heaven,
And Man, disgusted with God,
Turned towards Eve,
Things looked like falling apart.

But Crow Crow
Crow nailed them together,
Nailing heaven and earth together–

So man cried, but with God's voice.
And God bled, but with man's blood.

Then heaven and earth creaked at the joint

Which became gangrenous and stank–
A horror beyond redemption.

The agony did not diminish.

Man could not be man nor God God.

The agony

Grew.

Crow

Grinned

Crying: "This is my Creation,"

Flying the black flag of himself.

British poet Ted Hughes (1930–1998) wrote an acclaimed but unsettling book in 1970 titled *Crow.* It was a new kind of work, a mytho-poetical narrative, that was experimental in its time for the way it referenced folk legends and world mythologies and for the language Hughes employed, which is bold, harsh, and almost primitive (like a crow's caw). The narrative that runs through the collection of interlinked poems is opaque but extremely emotional. Rhythms inside the poems are chant-like and hallucinatory; words and phrases repeat over and over. And the central character is Crow, who is vulgar, disruptive, confused, anarchic but also a creator and funny at times. He is a classic Trickster, mischievous but not entirely evil. In this poem, Crow tries to repair Man's relationship with God by indelicately nailing them to one another (much as Christ was nailed to the cross). He grins at his creation but it is a horrible thing—doomed to a gangrenous death. Here, Crow has attempted to create something new using old parts but he has failed; probably because he does not really understand what it means to be Man or God, Earth or Heaven.

***Crow Ikon.*** **1982. Leonard Baskin. Lithograph printed in color on Velin d'Arches paper, 48 × 37.5 inches. Leonard Baskin's crows almost always took on human features. He portrayed dozens of crows in various states of disarray or disintegration, hinting at mankind's own state of existence. Here an agonized crow becomes an ikon, an object of macabre veneration (Gift of Philip and Dorothy Green, Smith College of Art, Northampton, Massachusetts, SC1992.30.9. ©Estate of Leonard Baskin).**

*Crow* came into being sometime after artist and friend Leonard Baskin (discussed in Chapter 10) asked Hughes to create poems for his crow drawings. Baskin's very human-like bird was arresting in its rough, disheveled features which made it look as if it was powerful but in pain. Hughes used this kind of crow as a mythic archetype and as a stand-in for the human impulse. Poet Jarold Ramsey writes: "Why did Hughes choose a crow as his Trickster-Transformer? In one sense the question is silly for anyone who has ever watched a real crow or raven for more than a few moments: if there had been no genus *Corvus*, Hughes or somebody would have had to invent it, as the saying goes, for its raucous intelligence, greediness, and blackness—an image of the human."[13] Hughes was very sensitive to the physicality of his poems, the way they looked on the page. He once commented: "In a way, I suppose, I think of poems as a sort of animal."[14] We can see here how he experimented with the graphic interplay of word as image. The words shred toward the end of the work. *Crow* put Ted Hughes at the top of modern poetic literature. He is recognized as one of the greatest poets of the twentieth century.

# 3

# Sky-Bound

The pleasure of watching a flock of birds dipping and diving over treetops on the way to a secret rendezvous fills me with sweet longing. If only we humans could join that flight in the free air over the earth's hard crust. If only we could wheel and soar and speed along an invisible line to somewhere else; just lift gravity from our shoulders and take off. Birds "on the wing" sadly leave us behind, burdened as we are by heavy bodies that must trudge and plod. Corvids do a spectacular job at forsaking us and as they disappear over the horizon, we are inclined to attach stories to them.

Strong, stiffly feathered wings allow the birds to travel long distances in order to locate needed resources. They also make it possible to play in the air with abandon. First-hand accounts of the acrobatics performed by crows and ravens are captivating. Naturalist David Rains Wallace tells us about ravens near his home in Northern California:

> They circled in the stiff eddies of wind above the knoll, hardly moving their wings, merely ruddering delicately with their wedge-shaped tails. At first, they didn't make a sound, just swept back and forth about 20 feet above me. They climbed, sideslipped, crossed each other's courses, swooped, and climbed again. It was mesmerizing to watch, as though the birds were weaving spells in the blue sky with their shapes like black crosses. They continued this silent weaving for several moments, and it evidently excited them. The pattern of their flight quickened. One bird croaked several times as though in exhilaration, and the four paired off. Each pair flew in unison, one partner close above the other as they made steep little dives. Pulling out of these dives, the birds made the same gurgling-drain sounds that I hear around my walnut trees. The pairing and diving seemed to complete the performance. Still hardly moving their wings, the ravens soared away on an updraft, vanishing as suddenly as they had appeared.[1]

## *Field Notes: Aerial Abilities*

Scientists have examined the exuberant aerial displays of crows and ravens in order to determine why the birds engage in the behavior. Ravens, in particular, are renowned for their maneuvers. They might choose an updraft to carry them two thousand feet into the air, then pitch straight down at velocity while executing rolls and swoops. Or one might perform a half-roll—folding a wing at the wrist, rolling quickly onto its back, then bending the other wrist and reversing direction. Pairs of ravens will carry out corkscrew dives, backward loops, and full barrel rolls in perfect tandem, wing tips touching. Four kinds of flight movements have been described for *Corvus corax*: the "loop flight," where a pair soars in circles over their territory; the "gliding and wave flight," where the pair flies in extremely close proximity low over treetops; the "dive," where the pair takes turns diving from a great height; and the "throwing onto the back" roll, where a bird flips its body. Dirk Van Vuren of the University of Kansas observed a raven near Santa Barbara, California, execute a single performance that included a sequence of 6 half-rolls, 2 full rolls, and 2 double-rolls.[2]

For both Dr. Lawrence Kilham and Dr. Bernd Heinrich, all this rising, twirling, and tumbling looks like tremendous fun, and both have documented specific flying behaviors that appear to be for pleasure alone. Dr. Heinrich, for instance, saved a letter from hang glider Tim Hall about his experience during a launch from a 3,000-foot cliff. Hall recounts that about ten ravens were swooping and barrel rolling around him when a new bird appeared with a long white plastic streamer similar to surveyor's tape in its beak. It began to sweep down through the others, fold its wings, then sweep back up again. After doing this a few times, it handed the streamer to another raven, who caught, dipped, then released it to the next. The ravens thus took turns playing with the object.[3]

Ravens soar more than crows do; crows never do somersaults. Nevertheless, flocks of sociable crows can perform beautiful acrobatics. These displays are most often associated with courtship—to draw attention to oneself or to demonstrate strength and ability to a potential mate—and we will look at that characteristic in Chapter 11. But crows can be simply goofy. On YouTube you can see a crow using the lid of a mayonnaise jar to slide down a snow-covered roof. A bird will also hang upside down from a wad of moss on a tree branch and joyously swing back and forth, or engage in a game of tug-of-war with an object, or take a stick into the air to drop and then recapture it before it lands. Crows will torment larger animals for no apparent reason. And, during a football game at the University of Washington, thousands of spectators watched as approximately fifty American crows played catch over the end zone with a crumpled ball of paper.[4]

Dr. Kilham described his pet Crowsy's delight in a pebble:

> One morning he picked up a pebble and, holding it in his bill, did a minuet, taking steps and sideways hops that reminded us of the dances of Crested Cranes we had watched in Africa. He also danced on a stone wall while holding bits of fresh leaves, stems, or flowers. If I returned from a walk and called, he would leave his spruce limb to alight by my feet, pick a small leaf, dance with it, then lay it by my foot or on my shoe.[5]

As already mentioned, ravens carry out "useless" playful behavior as well. They will "body surf" down snow-covered hillsides, dive in and out of irrigation waterspouts, and amuse themselves with "ground games." Naturalist Bob Armstrong recalls seeing ravens play a game in Juneau, Alaska. A bunch of ravens placed themselves in a circle. One at a time, a raven would pick up an object—a piece of foam, a wad of paper—and walk into the center of the circle. It would stop for a minute to let the rest see what it had, then return to its spot while another raven walked to the middle with another object. One raven found a large paper cup that it struggled to pick up and walk into the middle with. It paused there and that was the end of the game![6]

Why do these birds engage in play? Scientists believe that the ability to "play" is a manifestation of intelligence. Bernd Heinrich writes: "I suspect that play is almost literally like intelligence. Play is an acting out of options, among which the best can then be chosen, strengthened, or facilitated in the future. The difference is that with play, the options are played out overtly, not only in the mind."[7] Through their high jinks and aerial acrobatics, crows and ravens expose themselves to a wide variety of experiences that in turn help to sharpen their coordination, facilitate social skills, teach them about their surroundings, and satisfy their innate curiosity. They profit from it.

When we see nature's black aviators performing in the sky, we are filled with wonder. They appear to be able to reach the stars and access places unknown. For this reason, crows and ravens have been associated with celestial locations and with powerful gods living in the heavens. In Australia, for instance, Crow was responsible for unleashing the winds. The Babylonian god Adad used the raven to bring rain and storms to the dry land at the end of the summer season. The Indian goddess Dhumavati decided to travel the universe on a crow while the Akkadian storm-god Anzu chose to take the form of a raven in order to pursue his activities. In Europe, crow and raven symbols tend to be associated with night and the moon and the Corvus constellation is named after the bird. In a later chapter we examine their roles as messengers to and from the gods. All this information suggests that corvids have two homes; they bridge the gap between heaven and earth.

## The Sun, the Moon, and Crow Crowson

Russia. Collected by Alexsander Nikolaevich Afasášev; translated by Leonard A. Magnus.

Once upon a time there was an old man and an old woman who had three daughters. The old man went into the loft for some groats, and took them home, but there was a hole in the sack, and the groats were running and running out of the sack.

The old man went home, and the old woman asked, "Where are the groats?" But all the groats had dripped out.

So the old man went to collect them, and said, "If only the Sun would warm the grain, and the moon show its light on it, and Crow Crowson help me to get the groats, I would give my eldest daughter to the little Sun, and my middle daughter to the Moon, and my youngest to Crow Crowson." So the old man set to collecting the grain, and the Sun warmed it, and the Moon shone on it, and Vóron Vóronovich [Crow Crow-son] helped to collect the grain.

The old man came back home and said to the eldest daughter: "You must dress nicely and go out on the steps." So she dressed and went out on the steps. And the Sun laid hold of her. And he commanded the next daughter in the same way to dress herself finely and to stand on the steps. So she dressed herself up and went out, and the Moon seized and took away the second daughter. And he said to the third daughter, "Dress yourself prettily and stand on the steps." So she dressed herself prettily and stood on the steps, and Crow Crowson seized her and carried her away.

Then the old man said, "I think I might go and visit my sons-in-law." So he went to the Sun, and at last he arrived there.

The Sun asked him, "With what shall I regale you?"

"Oh, I don't wish for anything!"

So the Sun bade his wife make a custard ready. So the daughter prepared the custard; the Sun sat down in the middle of the floor, and his wife put the pan on him and the custard was soon cooked. So they gave the old father refreshment.

Then the old father went back home and bade his wife make him a custard; and he sat down on the floor and commanded her to put the pan with the custard on to him.

"What are you talking about? Bake it on you?!" said the old wife.

"Go on!" he replied. "Put it there; it will be baked!"

So she put the pan on him, and the custard stood there for ages and was not ever cooked, only turned sour. It was no good. So in the end the wife put the pan into the stove, and this time the custard was baked and the old man got something to eat.

Next day the old man went to stay as a guest with his second son-in-law, the Moon, and he arrived.

And the Moon said, "With what shall I regale you?"

"I do not wish for anything," said the old man.

So the Moon got the bath heated ready for him.

The old man said, "Won't it be very dark in the bath?"

"No," said the Moon to him, "quite light; only step in."

So the old man went into the bath, and the Moon twisted his little finger into a chink, and it was quite light in the bathroom. So the old man steamed himself thoroughly, went back home and told his wife to heat the bath at night. So the old woman heated it, and he sent her there to steam herself.

"But," she said, "it will be much too dark to steam myself!"

"Go along! it will be light enough."

So the old woman went. And the old man saw how the Moon had lit the place up for him, and he went and bored a tiny hole in the bathroom and thrust his finger through it.

But there was still no light in the bath, and the old woman shrieked out to him, "Dark! much too dark!" It was not any good. So she went out, brought a lamp, and enjoyed her steam bath.

On the third day the old man went to Vóron Vóronovich. He got there.

"How shall I regale you?" asked Vóron Vóronovich.

"Oh," said the old man, "I don't want anything!"

"Well, let us come and sleep on the perch."

So the Crow put a ladder up and climbed up there with his father-in-law. Crow Crowson settled himself comfortably with his head under his wing. But as soon as ever the old man dropped off to sleep both of them fell down and were killed.

The story of the Sun, the Moon, and Crow Crowson (also known as Vóron Vóronovich) is intriguing because its surprise ending is equally tragic and funny. We encounter a hapless human whom we follow as he casually visits powerful beings in their homes, only to lose his life at the hands of an incompetent crow. Translator Leonard Magnus noted that Russian folktales feature down-to-earth settings and perfunctory reporting of ill-fated events. No moral lessons are offered, only a sense of kinship with the main character and of being transported to other worlds.

The crow appears repeatedly in Russian lore, where it is not unusual to marry one's child to the bird. The crow is associated with potent forces (such as the sun and moon) and is invoked in magical charms. One such charm was recited by Russian soldiers to protect themselves on their way to war: "Do thou bid, by my enchanting words, the crow to gain me the seven *pud* key. The crow has smitten the house of bronze, has pecked the fiery snake to death, has gained the seven *pud* key. With that key I will unlock the princely castle, … I will gain the knightly gear, … and in that gear the arquebus cannot fell me, the shots cannot hit me, the warriors and champions, the hosts of Tatary and Kazán cannot hurt me."[8] We can only hope the charm worked.

Alexsander Nikolaevich Afasášev (1826–1871) trained as a lawyer in Moscow but was drawn to his country's folkloric tradition. He collected over 600 tales for the Russian Geographic Society and, although he died penniless at the age of 45, his contribution to our understanding of Russian storytelling remains a great legacy.

## A Living Corpse

Yakut (Sakha), Siberia. Told by Vasilij Popov, Nasleg of Chadan, Ulus of Chotschin. Collected by G.V. Ksenofontov; translated into English by Stefan Neil Klemenc.

As the living dead came to his senses it was dark, as the night had already fallen. Shortly after that a black raven swooped down, put its head between the legs of the man, lifted him up, and flew up with him. Up there was an opening. They flew right through it to a place. There, both the sun and the moon were shining and the houses and barns were made of iron. The people there had the heads of ravens and bodies like humans.

This scene takes place in the middle of a long Yakut myth that features a shaman named Aadja who dies, is raised up out from his coffin, taken on many adventures in other worlds, and eventually reborn into this world. It was told to G.V. Ksenofontov by Ivan Popov on March 22, 1925.

Shamans (both men and women) have historically played an essential role in Arctic cultures. They were able to cross the boundaries of existence to uncover the causes of illness. They figured out what invisible elements were affecting a community and had influence over both the weather and the appearance of game animals. They were healers, seers, mediators, guardians, and leaders in their communities. Stories of shamanic activities abound in the polar regions of the North. And a raven was usually present. Waldemar Bogoras reported that the Chukchee of Siberia considered the raven to be an "assistant" to the Creator (or the Spirit of the Zenith) and thus it received part of all sacrifices directed upward.[9] Associating with the raven brought shamans enhanced powers. Scholar Jarich Oosten describes the raven as a:

> master of light as well as the alternation of light and dark that is at the core of the raven complex in the western Arctic. Light acquired a special importance in Inuit traditions in the notion of qaumaniq [vision] and light. In order to become an angakkuq [shaman] a person had to obtain qaumaniq.... The stronger the light is within him, the deeper and farther away he can see, and the greater is his supernatural power. The shamanic vision enabled the angakkuq to see hidden things such as spirits and shades, or transgressions of tirigusuusiit [things people had to abstain from].[10]

In this story, we are introduced to raven-headed spirit-people who live in the upper branches of the giant larch tree that reaches into the heavens. Ravens were said to fly easily between this home and the Middle World of men. G.V. Ksenofontov (1888–1938), a specialist in the ethnography and folklore of the Yakut, Buryat and Evenk peoples, was able to capture the words of these remote dwellers of the Arctic north.

## The Origin of Daylight

Tahltan, British Columbia. Collected and translated by James A. Teit.

At this time there was no daylight, or sun, moon, or stars. Raven went to a village and asked the people if they could see anything. They said, "No, but one man has daylight, which he keeps in a box in his house. When he takes off the lid, there is bright light in his house." The people could not work much, for it was night continually. Raven found out where Daylight-man lived, and went to his house. This man also had control of the sun, moon, and stars. Raven went into the house and came out again. He planned what to do to get daylight for himself and the people.

Daylight-Man had many slaves, and a daughter who had been a woman for three years, but she was still undergoing the ceremonies encumbent on girls at puberty. She lived apart in the corner of the house, in a room of her own, and was closely watched. She drank out of a white bucket every day, and she always examined the water before drinking, to see if there was anything in it. Slaves always brought the water to her.

> Raven changed himself into a cedar-leaf in the bucket of water the slave was bringing. The girl noticed it, and before drinking threw it out. He assumed his natural form again. Next day he transformed himself into a very small cedar-leaf, and hid in the water. The girl looked in the water, and, seeing nothing, she drank it all, and thus swallowed Raven.... After nine months she gave birth to a son. Her parents said they would rear the boy and acknowledge him as their grandson, even if he had no father....
>
> The boy grew very fast, and soon was able to walk and talk. His grandfather loved him dearly. One day he cried very much and wanted to be allowed to play with the moon. His grandfather ordered the moon to be taken down and given to him. The boy was pleased, and played with it until tired; and then they hung it up again. After a while he got tired of the moon and cried much, saying he wanted the sun. It was given to him; and he played with it until tired, then gave it back, and the people hung it up again. After a while he became tired of the sun, and cried for the Dipper (stars). Now they allowed him to play with these things whenever he wanted. After a long time, when he felt strong, he cried for the daylight. His grandfather was afraid to give it to him, because it shed so much light; besides, whenever it was lifted up, the sun, moon, stars, and everything worked in unison with it. It was their chief. At last, however, the boy was allowed to have the daylight, but his grandfather was uneasy when he played with it. When the boy lifted up daylight, much light would come; and the higher he held it, the brighter became the daylight. On these occasions when the boy held the daylight high, the old man would say, "Eh, eh!" as if he was hurt or extremely anxious. The boy balanced the daylight in his hands to get used to carrying it.
>
> At last one day, he felt strong enough for the feat he intended to perform. He put two of the toys in each hand and balanced them. He felt he could carry them easily. Then, at a moment when the people were not watching, he flew out of the smoke-hole with them. He threw daylight away, saying, "Henceforth there shall be daylight, and people will be able to see and work and travel. After dawn the sun will rise; and when it sets, night will come. People will then rest and sleep, for it will not be easy to work and travel. Then the Dipper and moon will travel and give light. These things shall never again belong to one man, nor be kept locked up in one place. They shall be for the use and benefit of all people." He threw daylight to the north, the sun to the east, the moon to the west, and the Dipper to the south. Since the introduction of daylight, people and game rise with daylight, and go to sleep with nightfall.

We return to James Teit's collection of Tahltan Raven tales. In Chapter 2 we learned how Raven managed, somewhat by force, to regulate the ebb and flow of the tides. Now we see how he was able to release daylight into the world. The myth of the origin of daylight is one of the most beloved stories told by Indigenous peoples

**Lkaayaak yeil s'aaxw (Box of Daylight Raven Hat). ca. 1850. Gaanax'adi Clan, Tlingit. Maple, mirror, abalone shell, bird skin, paint, sea lion whiskers, copper, leather, Flicker feathers, 11.87 × 7.75 × 12.25 inches. This hat, depicting Raven in a semi-human form as he grasps the lid of the box that contains the sun, moon and stars, would have been a most valued object that was brought out only on ceremonial occasions of great importance. It was associated with a particular clan and when it appeared the history and connection to the clan would be recounted. Similar hats and regalia are still used during important occasions (Seattle Art Museum, 91.1.124, Gift of John H. Hauberg. Photo: Paul Macapia).**

along the Northwest coast of Canada and the United States. In hundreds of versions there is a box of daylight that is opened through a bit of wily thievery. In some stories, daylight is held within a series of nesting boxes and Raven-child has to convince his grandfather to open the boxes one by one until, from the smallest one which contains all the light in the universe, an incandescent ball is brought out that he quickly snatches up. Raven is a master trickster and we will read about some of his further exploits in later chapters, but Raven is quick to remind us: "I have fooled people, but always so that good should come. I have made things easier and better for the people."[11]

## The Bringing of the Light by Raven

Told by Paimut. Bering Sea Eskimo (Inuit), Lower Yukon, Alaska.
Collected and translated by Edward W. Nelson.

In the first days there was light from the sun and the moon as we now have it. Then the sun and the moon were taken away, and people were left on the earth for a long time with no light but the shining of the stars. The shamans made their strongest charms to no purpose, for the darkness of night continued.

In a village of the lower Yukon there lived an orphan boy who always sat upon the bench with the humble people over the entrance way in the *kashim.*[*] The other people thought he was foolish, and he was despised and ill-treated by everyone. After the shamans had tried very hard to bring back the sun and the moon but failed, the boy began to mock them, saying, "What fine shamans you must be, not to be able to bring back the light, when even I can do it."

At this the shamans became very angry and beat him and drove him out of the kashim. This poor orphan was like any other boy until he put on a black coat which he had, when he changed into a raven, preserving this form until he took off the coat again.

When the shamans drove the boy out of the kashim, he went to the house of his aunt in the village and told her what he had said to them and how they had beaten him and driven him out of the kashim. Then he said he wished her to tell him where the sun and the moon had gone, for he wished to go after them.

She denied that she knew where they were hidden, but the boy said, "I am sure you know where they are, for look at what a finely sewed coat you wear, and you could not see to sew it in that way if you did not know where the light is." After a long time he prevailed upon his aunt, and she said to him, "Well, if you wish to find the light you must take your snowshoes and go far to the south, to the place you will know when you get there."

The Raven boy at once took his snowshoes and set off for the south. For many days he traveled, and the darkness was always the same. When he had gone a very long way he saw far in front of him a ray of light, and then he felt encouraged. As he hurried on the light showed again, plainer than before, and then vanished and appeared at intervals. At last he came to a large hill, one side of which was in a bright light while the other appeared in the blackness of night. In front of him and close to the hill the boy saw a hut with a man nearby who was shoveling snow from the front of it.

The man was tossing the snow high in the air, and each time that he did this the light became obscured, thus causing the alternations of light and darkness which the boy had seen as he approached. Close beside the house he saw the light he had come in search of, looking like a large ball of fire. Then the boy stopped and began to plan how to secure the light and the shovel from the man.

After a time he walked up to the man and said, "Why are you throwing up the snow and hiding the light from our village?" The man stopped, looked up, and said, "I am only cleaning away the snow from my door; I am not hiding the light. But who are you, and whence did you come?" asked the man. "It is so dark at our village that I did not like to live there, so I came here to live with you," said the boy. "What, all the time?" asked

---

* A *kashim* is a building used by Inuit as a community gathering place.

the man. "Yes," replied the boy. The man then said, "It is well; come into the house with me," and he dropped his shovel on the ground, and, stooping down, led the way through the underground passage into the house, letting the curtain fall in front of the door as he passed, thinking the boy was close behind him.

The moment the door flap fell behind the man as he entered, the boy caught up the ball of light and put it in the turned up flap of his fur coat in front; then, catching up the shovel in one hand, he fled away to the north, running until his feet became tired; then by means of his magic coat he changed into a raven and flew as fast as his wings would carry him. Behind he heard the frightful shrieks and cries of the old man, following fast in pursuit. When the old man saw that he could not overtake the Raven he cried out, "Never mind; you may keep the light, but give me my shovel."

To this the boy answered, "No; you made our village dark and you can not have your shovel," and Raven flew off, leaving him. As Raven traveled to his home he broke off a piece of the light and threw it away, thus making day. Then he went on for a long time in darkness and then threw out another piece of light, making it day again. This he continued to do at intervals until he reached the outside of the kashim in his own village, when he threw away the last piece. Then he went into the kashim and said, "Now, you good-for-nothing shamans, you see I have brought back the light, and it will be light and then dark so as to make day and night," and the shamans could not answer him.

After this the Raven boy went out upon the ice, for his home was on the seacoast, and a great wind arose, drifting him with the ice across the sea to the land on the other shore. There he found a village of people and took a wife from among them, living with her people until he had three daughters and four sons. In time he became very old and told his children how he had come to their country, and after telling them that they must go to the land whence he came, he died.

Raven's children then went away as he had directed them, and finally they came to their father's land. There they became ravens, and their descendants afterward forgot how to change themselves into people and so have continued to be ravens to this day.

At Raven's village day and night follow each other as he told them it would, and the length of each was unequal, as sometimes Raven traveled a long time without throwing out any light and again he threw out the light at frequent intervals, so that the nights were very short, and thus they have continued.

Storytelling was a favorite pastime among the peoples Edward W. Nelson (1855–1934) visited. Little girls often gathered together to tell "knife stories" which they illustrated using their knives to draw lines on the ground. At night, elders assembled in the kashim (or, *qasgiq*) to regale communities with myths (told in low, dramatic tones) about the ways of the ancestors and the beginnings of all things. The people also created mythologies to reflect how unseen forces were still affecting their lives. "The Bringing of the Light by Raven" might be this kind of tale for it explains the origin of the dramatic changes in the amount of daylight that is experienced in the area over the course of a year. Nelson noted that young men who had an aptitude for learning tales became narrators and repeated them verbatim, even with the accompanying inflections of voice and gestures.

The communities that Nelson studied were identified as Bering Sea Eskimo. By today's standards the term "Eskimo" is no longer considered appropriate. It is a term that was applied to northern peoples by Westerners. They did not call themselves Eskimo. Confusingly, peoples in this area did not call themselves "Inuit" either. They each spoke a distinct language and might be better identified as Aleut, Yupik, or Iñupiat, but these terms are not entirely accurate either. Here, they are identified as Inuit to denote their Indigenous claim to Northern Alaska, Canada, and Greenland.

Inuit communities in the lowland regions of western Alaska, between the Bering Strait and the Aleutian Islands, were among the least known in the world. Ignored by explorers, missionaries and whalers alike, yet comprising one quarter of the entire Inuit population, the Bering Sea people lived in large stable villages with substantial homes and a plentiful supply of sea food.

Of their mythic Raven, Edward Nelson wrote: Raven Father "came from the sky ... [and] is believed still to live there.... The Raven Father, who made the land and everything upon it, is the subject of many tales in which he is represented as benefiting mankind. When he returned to the sky he left on earth children like himself, ... and while the ravens now living are thought to be descendants of the Raven Father they have lost their magical powers."[12] To read about Edward W. Nelson's work with the native people of the Bering Sea, turn to Chapter 10, Transformation.

## Crow that Saved the Sun

Ainu, Japan. Collected and translated by the Rev. John Batchelor.

Version One:

When God [known as the Great Kamui] created the world an evil one did all he could to frustrate the designs. After all things were made, he perceived that mankind could not live without the warmth and light of the sun. He therefore made up his mind to destroy that beautiful and useful work of creation. So he got up early one morning, before the sun had risen, with the intention of swallowing it. But the Great Kamui was aware of his designs and made a crow to circumvent them. When the sun was rising the evil one opened his mouth to swallow it; but the crow, who was lying in wait, flew down his throat, and so saved it.

Hence the crows, remembering how they saved the human race,

now have the idea that they can be just like men and live upon the food provided for the sustenance of their families. We find that they have good cause for being bold and saucy, and it is not for us to say that crows are useless creatures.

Version Two:

A very long time ago, when the sun was about to rise for the first time, the evil one opened his mouth to swallow it. Upon this a multitude of crows—they being the most numerous birds in the world—flew down his throat. This disconcerted him so much that while he was busily engaged in getting rid of the feathers out of his mouth the sun arose, and it became daylight. And so the Ainu were all able to hunt and fish, and did not perish, as the demon desired. For this reason, then, crows must be borne with and not grumbled about, even though they are bold and saucy and take the food away from the people. And although they should not be prayed to, *inao* (religious objects) may be offered to them.

The Rev. John Batchelor (1855–1944) lived for twenty-five years among the Ainu people of northern Japan, an Indigenous group whose rich culture has sadly just about disappeared. He is representative of many Christian missionaries who fanned out across the globe during the early part of the twentieth century to proselytize. While the effects of their presence can legitimately be criticized, in many cases these missionaries were responsible for collecting information that is now enlightening both researchers, historians, and native peoples alike. Batchelor sharply and publicly criticized the Japanese government for its treatment of the Ainu, which was very similar to U.S. and Canadian treatment of native groups. Ainu were forbidden to hunt, speak Ainu, obtain an education, or practice their traditions. They were forcibly segregated into small villages. His book, *The Ainu of Japan*, offers us a glimpse into the lives of this vanishing people.

Both versions of the story about how crows saved the sun contain a gentle humor. We are admonished to treat the natural grain and food robbers kindly because we owe them a great debt. Batchelor also recorded the "Legend of the Rook" which says: "Very long ago, when the Ainu saw rooks for the first time, they thought that as they descended from the heights above and were clothed in such beautiful and glossy dresses they must be gods." Hunters would sacrifice a rook in ritual honor; in return their hunts were expected to be successful.[13]

A more developed association of the crow with the sun appears in Japan's earliest written works where Yatagarasu ("eight span crow" meaning really large crow) appears as a holy bird and servant to the sun goddess, Amaterasu. The *Kojiki* ("Records of Ancient Matters," seventh century CE) recounts how Yatagarasu, an excellent crow-navigator, was able to lead Japan's first emperor Jimmu through the mountains on his march from the western lands to Yamamoto to establish his country.[14] Yatagarasu has also been identified as the "raven of many feet." By the 930s CE its mythology had

**Kumano Hongu Taisha shrine. Japan. The Kumano area of Japan centers around three shrines, all of which pay tribute to the three-legged crow who helped Japan's first emperor Jimmu establish the country. Pilgrims have traveled to the shrines for over one thousand years (photograph courtesy of Sydney Solis).**

merged with the Chinese myth of the three-legged sun crow (which we read about in the following story). The sacred bird is now prayed to at various shrines around Japan.

## The Great Archer Yi Shoots Down Nine Suns

Ancient China. Retold by Jan Walls and Yvonne Walls.

[There once were ten suns]. The ten suns were in fact the ten sons of *Di Jun* (Lord Superior) who was said to have had two wives. One of them, the Goddess of the Moon, *Chang Xi*, gave birth to the twelve moons.

The other was the Goddess of the Sun, *Xi He*, who gave birth to the ten suns.

The ten suns lived in the *Tang* (Hot Water) Valley to the southeast, beyond the sea. Since the suns often bathed in this valley, the water there was boiling hot. There was a giant mulberry tree growing in the center of the sea near the Tang Valley called *Fusang* (Supporting Mulberry). This tree was over the thousand feet tall and more than one thousand arm-spans thick. The ten suns dwelled in this tree: nine on top and one at the bottom; and they took turns going out on duty in the sky. Therefore, even though there were ten suns altogether, people on earth never saw more than one at a time.

Every morning, one of the suns would take a bath in the *Xian* (Encompassing) Pond of the Tang Valley, then climb from the bottom of the mulberry tree to the very top where it would mount a carriage drawn by six dragons and driven by their mother Xi He to traverse the sky. Xi He would accompany one of her sons all the way to the end of the journey at the Spring of Sadness. There, she would stop the carriage to wait for her son's rays to fade away and her son to descend into the Yu Deep. Then she would drive the empty carriage back to the Tang Valley to pick up the next sun for the next day's journey. Thus it was that the ten suns would be accompanied by their mother to go on duty in an orderly way each and every day.

One day during the time of the sage king Yao—no one knew exactly how it came about—the ten suns suddenly refused to go on duty in an orderly fashion. All at once, all ten suns leaped out together and ran wildly through the sky, having great fun among themselves. This naturally brought an abundance of light to the earth, but it also brought great disaster.

The suns' brilliant rays scorched the earth and dried up all the grains that had been planted in the soil; rocks and metals were on the verge of melting, and people were so hot they could hardly breathe. Food became scarce and famine was widespread. As if this were not enough, monsters and fierce beasts came out of the forests and swamps, bringing even more suffering to humankind until finally the very survival of the species was threatened.

Di Jun and Xi He tried without success to persuade their sons to behave themselves while at the same time, King Yao was pleading to Di Jun to help deliver his people from their misery. Finally, Di Jun had to send the great archer, Yi, down to earth to slay the monsters and beasts and try to curb the misbehavior of the suns. Di Jun gave Yi a red bow and a bag of white arrows with which to carry out his mission, and Yi left for earth accompanied by his wife Chang E.

The main source of Yao's worries and the people's misery was, of course, the ten suns. Everyone begged Yi to do something about them. Yi first pretended to shoot the suns, hoping that he might frighten them away. The suns, however, were not the least bit intimidated. This so angered Yi that he walked to the center of a field, drew his bow to the fullest, aimed at one of the suns and with a swish, let an arrow fly up into the sky. A moment later, a ball of fire silently exploded in the air, flames darted in all directions and golden feathers came fluttering down to earth. Finally, a red bright object fell to the ground with a thud, and everyone rushed forward to discover a huge raven with three legs. This was a transformation of the spirit of the sun, of which there were now only nine left in the sky. Yi sent arrows flying into the sky one after another. Fire balls exploded, great sparks flew in all directions and feathers fluttered down in great profusion as the three-legged ravens fell down one by one. King Yao, thinking that one sun could still do great service for the people, asked Yi to leave the last sun unharmed.

Nine three-legged ravens had fallen to the ground. But where did all the fire balls go? It was said that they all fell into the vast ocean east of the sea to from a giant rock forty thousand *li* (about 13,000 miles) thick and forty thousand *li* in circumference. It was called "Wo Jiao" (The Fertile Scorch). Sea waters that dashed upon it would evaporate and disappear instantly. This is the main reason why, even though all the waters of all the rivers and streams empty into the sea, the sea never overflows.

So, a sun-raven (*jīwū*) lives within the sun as a dynamic force. What an amazing thought. The story of Archer Yi shooting down the suns is known to nearly all Chinese people today, but its roots are ancient, dating as far back as the fifth century BCE. Chinese mythology exists in fragments and only from gathering its scattered pieces can one put together the myth of the three-legged raven. And it is quite a story: ten ravens (often called crows) ushering ten suns across the sky until chaos breaks out and Archer Yi must come to humanity's rescue. It is the story of a great disruption and the restoration of the natural order of things.

The motif of a bird carrying the sun across the sky, or settling in the sun, appears in imagery as far back as the middle to late Neolithic period (ca. 7000–1700 BCE). The first mention in literature comes in the form of a "Heavenly Question" posed by the poet Ch'ü Yüan (ca. 340–278 BCE), who asks "Why did Yi shoot down the suns?/Why did crow feathers fall from them?" The later Eastern Han commentator Wang Chong (25–100 CE) substitutes "three-legged" for *jun* (ruler/god) and Gao Yu (ca. 168–212 CE) also identifies the *jun*-raven as three-legged. The idea of the birds living in a Mulberry Tree certainly predates Ch'ü Yüan's time as evidenced by imagery on a fifth century BCE lacquered chest.[15]

In our story, the raven is identified as the spirit of the sun. Some modern scholars theorize that early observation of sunspots might have inspired the fabrication of a solar crow.[16] We will never know. What we do know is that the raven held a powerful role. The bird was associated with one the most essential elements of our lives, the sun.

*Above and below:* Lacquered suitcase with painting of Hou Yi. ca. 433 BCE. Leigudun tomb (No. 1), Suizhou, Hubei, China. 27 × 19.25 × 14.5 inches. Unearthed in 1978 from the tomb of Marquis Yi of Zeng, this container is painted with the story of Archer Yi. A tiny archer in the far-right corner of the chest has just shot a bird who falls from a tree (Photograph courtesy of the Hubei Provincial Museum, Wuhan, China; drawing Hubei Bowuguan and Beijing gongyi meishu yanjiusuo [Ed.]. 1984. *Zhanguo Zeng Hou Yi mu chutu wenwu tu'an zuan*).

## What the Crow Said

Michael Hannon, USA.

Though friendly to magic
I am not a man disguised as a crow.

I am night eating the sun.

Correspondences attach themselves to crows—magic, disguise, nighttime, the sun. There is a primal, elemental nature to the crow in Michael Hannon's poem. The

***Moon Mad Crow in the Surf.*** **1943. Morris Graves. Gouache on paper, 23.5 × 28 inches. The artist once said "I paint to evolve a changing language of symbols, a language with which to remark upon the qualities of our mysterious capacities which direct us toward ultimate reality" (Miller, 1942). The crow in this work appears as a stand-in for our yearning to reach the ultimate reality he refers to (Brooklyn Museum, Gift of Mrs. Milton Lowenthal, 1991.109.2. © Morris Graves Foundation).**

bird is a force, not a pretender; and while admittedly magical it is more powerful than magical. What will happen to us when night eats the sun?

Michael Hannon (b. 1939) has been writing spare, penetrating poetry for more than fifty years. The poet explores the mystery of nature as well as the psyche of humankind. Enigmas attract Hannon's attention, and revelation is his work. He has said: "My poetry has always been a philosophical investigation of that mystery."[17]

Artist Morris Graves (1910–2001) seems to have felt similarly. Spiritual in the same vein as Hannon, Graves was deeply connected to his natural surroundings in remote Fidalgo Island on the Puget Sound, Washington, and he painted the sights and sounds around him to symbolically express the inner human spirit. In *Moon Mad Crow in the Surf,* we see the light of the Northwest coast as a gauzy glow. It has bewitched the crow who seems to be physically pulled out of the water. The crow is fixated on the moon; its sharp black beak and piercing eye are aimed at the celestial object. Black is drawn up to white. Graves was known to roam at night and paint until morning. He listened intensely to the sounds of the night, trying to visualize and then draw them. At the same time, he infused the natural world with a sense of terrible longing. Anxiety permeates this work, as though it is our own psyche that may have gone mad. *Moon Mad Crow in the Surf* seems to reflect the war time period during which it was painted. It feels like an attempt to escape the earthbound world for a higher realm.

# 4

# Language and Song

In a distant dream-like time, Man and Nature were once indivisible and humans could communicate easily with the natural world. All creatures could speak a common language, and they understood a common universe. We have often longed to return to this time. We find, through myth and legend, that we are able to travel back and reclaim the connection we once had. We hear again the stories of spirit beings. We heed their wisdom and relive their humorous exploits.

The crow and the raven feature prominently in myths about language because they seem to represent for us our own frustrating condition. Crows and ravens are oxymorons—birds that can't sing. They have somehow yielded the distinctive language of bird song, just as we have surrendered our ability to converse with animals. They become for us an embodiment of our own plight in nature. Some cultures celebrate the variety of crow and raven vocalizations. Others (as demonstrated in a well-known Korean quip: "Crow has twelve notes, none of them music") practically dismiss the birds from the avian kingdom. And then we find earnest treatises on the precise meanings of raven or crow pronouncements, for some peoples believe those unintelligible *quorks* and caws portend something important.

When I spend time with the crows that hang around my house, I hear the occasional chortle, the sharp snap, or repeater pops sent into the air. But there is no translation manual; I am left mystified about the meaning. It is unfortunate that humankind can no longer understand the thoughts of a crow and raven, for they do indeed communicate—vociferously. They broadcast who they are, how they feel, what they want to do, and why they want to do it. These utterances may not sound like any kind of song, but they serve the same function.

## *Field Notes: Vocalizations*

The number of calls is astounding. A professor at the University of Bern in Switzerland identified and analyzed 81 distinct raven calls.[1] He reported that some of these were area-specific, representing "dialects" of raven language. Of the American crow, Lawrence Kilham reminds us that vocalizations depend on the individual and its sexual maturity, the breeding group, and the territory. Kilham adds: "It is impossible to describe the mixtures of *cohs*, caas, moans and growls, mingled with caws, low *cu-koos*, *g-wal-ops*, and kuck-woos, given at rates of up to sixty a minute that make up the 'songs' of crows."[2]

Corvid calls are most commonly established by convention and learned by the group. But the birds are capable of being creative. Ornithologist John Marzluff writes: "Most corvids quickly invent new vocal symbols and use them in appropriate and novel situations."[3] Crow sounds are known to vary in duration, pitch, harmonic structure, and vibrato. The birds will string together a changing series of calls to communicate specific information. Marzluff reports an experiment where they were able to rearrange a recorded sequence of crow caws to attract crows on the one hand and repel them on the other.[4] Different "sentences" convey different things and the birds will then clarify their rich array of vocalizations with postures, gestures, and facial expressions. Scientist Bernd Heinrich states that "even though we have little or no idea what it says, the raven has a greater variety of call than perhaps any other animal in the world except human beings."[5] Marzluff adds that "most corvid vocal communication systems meet at least the basic standards of a language."[6]

***The Concourse of the Birds*. ca. 1610. Habibillah of Sava. Isfahan, Iran. Ink, opaque watercolor, gold, and silver on paper. Folio 11r from the manuscript *Mantiq al-Tayr* (Language of the Birds) by Attar of Nishapur (Farid al-Din Attar). 10 × 4.5 inches. Originally written in the 12th-century, the Sufi allegory describes a discussion among birds about whether or not to follow the hoopoe to its divine king's home. The birds present their excuses and, out of 100,000, only 30 reach the goal. At center-right, we see that the crow is in the mix, head turned at the approach of a man (Fletcher Fund, 1963. Metropolitan Museum of Art).**

So, what do they typically "talk" about? Interestingly, crows generally do not "sing-call" to attract mates, using instead soft sounds, bowing postures, and mutual caresses to make themselves appealing. In other situations, they will employ scolding calls, dispersal calls, threat calls, contact calls, hunger calls, or feeding calls. For example, when a crow notices a great horned owl perched on a tree in the daytime, it will give an "assembly" call and every crow within hearing distance will hurry to the area to mob and harass that owl out of range. Researchers have also witnessed the ability of young ravens to recruit one another's help at the site of a fresh carcass by broadcasting unique "yells." They might also coordinate their actions to draw another animal, like a wolf, to open up the carcass.

Structured territorial calls, with pauses left for response, might fill the daytime air while at night one hears family members greet each other. Dr. Cyndy Parr of the University of Michigan has noted that "there are many soft sounds crows make when talking to family members, including rattles, growls, gargles, coos, squawks, squeals and plaintive oo-oo's."[7] If two members of a group are particularly friendly, they will adjust their "songs" to resemble each other more closely. Ravens have the ability to recognize the voices of friends from the past. Researchers recorded the vocalizations of a group of birds who were then separated. After three years apart, the scientists played the recordings to individual ravens. Amazingly, the ravens answered—with their "friend" voice.[8] They had not forgotten their old buddies.

Crows and ravens also communicate with humans. They can broadcast their anger with us very clearly. Ravens give deep, long, rasping caws and back up those declarations with a hammering against wood, or by throwing items, or through a pitched dive. In captivity, corvids have been able to mimic human language the way parrots do. Their complex throat muscles allow them to verbalize phrases. Animal behaviorist Konrad Lorenz's pet raven, Roah, used to utter his own name in a "human"' voice when he wanted to call for Lorenz. Lorenz writes: "Roah is, so far as I know, the only animal that has ever spoken a human word to a man in its right context...."[9] Scientists acknowledge that imitating human speech is an advanced cognitive trick that involves both learning and thinking.

Writer Charles Russell recounts a beautiful conversation with a raven:

> On occasion I had observed a pair of ravens on the beach in front of camp.... I began to vocalize with one of these birds, which was picking through seaweed and flotsam about 50 feet from where I was replacing the spark plugs in our Zodiac's outboard motor. I started the exchange by mimicking his clucking. When I saw he was interested, I put down my tools and, to amuse myself, began asking questions about the legend of the Guarantee. He responded by jumping up onto a rock that was part of an ancient fishing weir that natives had used to corral salmon as the tide receded.... Perhaps I'd been alone too long, but it seemed as though my raven friend had been waiting for an opportunity to talk. He babbled away for several minutes. Ravens have a tremendous variety of vocalizations in their repertoire, and this bird's long dissertation in answer to my questions included some very strange hooting noises and guttural tongue rolls.... Our conversation went on for twenty minutes ... as I was making my last trip between camp and boat, I heard a raven cluck at me from the top of a large boulder only 10 feet from the trail. It was the bird I had talked with a couple of days before, and he seemed to want to pick up our conversation where we'd left off. I was surprised at how gentle he appeared to be. I had never before been this close to a wild raven. He was fluffing his feathers while softly doing his tongue roll. When I tried taking a couple of steps toward the rock, he stayed where he was, ... Moving my wrist and fingers in a slow stroking manner, I reached carefully out to him as I softly talked about the hazards of boating and flying in the fog. The raven was quiet but had hunched his shoulders and was holding his head low. His wings were stretched out along the top of the rock.[10]

Bernd Heinrich notes: "We know infinitely less about vocal communication in ravens than we know about the call of a frog, a cricket, or the zebra finch."[11] And, sometimes we just don't know why they make the sounds they do. Lawrence Kilham has said that, sometimes, it seems "as if the performers were practicing or playing with their voices."[12] Corvids have been known to mimic car engines, toilets flushing, and radio static. A curator at an Idaho zoo said that their resident raven, who lived next to ducks, learned to quack, and would chatter away and then laugh "haw-haw-haw" as if he loved his own jokes.[13]

Whatever they are saying, corvids say it well. Bernd Heinrich remembered a time when a raven vocalized without any fellow bird around to hear him:

> Number 34 remained,.... For an hour, he sang. His song was so uplifting and exuberant to my ears that I got out my tape recorder and started recording. I sat down less than fifteen feet from him and he paid me no attention. He was gurgling, chortling, yelling, trilling, bill-snapping, quorking, and making sounds like water rattling pebbles.... As he sang, he raised his head high, often turning and gazing in all directions, alternately preening, stretching, picking at twigs, and gulping bills full of snow. I talked to him, telling him how beautiful his raven song was, that it was the most beautiful thing I'd ever heard.[14]

## *A Bird Artist: Kawanabe Kyōsai*

Traditional Japanese painting generally identifies itself through bold outlines, limited colors, flat space, and a sensitivity to the natural world. We see in *Crow on a Branch* a lunging dip, an open-mouthed beak, and a fierce look that all signal an intentional moment in the life of an ordinary crow. Japanese artist Kawanbe Kyōsai (1831–1889) brilliantly captures the bird in action. We can almost feel the grip of talons and rustle of puffing feathers. This is a bird with something on its mind. Kawanabe paired the clean, powerful outline of the dark bird with softer half-tone shades for the ink outline of tree branch and foliage. Softer still are the clouds of mist that fill up the scene. All eyes are on the action of the bird whose opening wings cross diagonally against the tree to form a compositional "X." We can imagine the sound of the warning "caw" that will penetrate the thick air. Kawanbe Kyōsai became known as "the crow artist" in his lifetime and his mastery is fully evident in his various crow works. The crow's realistic posture originated from Kawanabe's observations in nature. His skill was derived from years of study in traditional *ukiyo-e* and Kanō methods. Crows were Kawanabe's signature subject and the creature he most frequently depicted. He even personally identified with the bird, using it as a motif for his seals.

Kawanabe Kyōsai was a boisterous man living in a tumultuous time. In the mid–1800s, Japan experienced civil war as it transitioned from a traditional political and military "shogunate" system to a more modern Meiji era administration. New systems and cultures, particularly Western cultures, were introduced that threatened familiar values and ways of life. The artist himself pushed boundaries. Temperamentally larger than life, Kawanabe was prolific, but he was prone to excessive drinking, eccentric work habits, and outrageous behavior. The works that best reflected his character were pieces filled with satire, bawdy humor, macabre subjects, and irreverent caricatures. Kawanabe went so far as to rename himself. He was born Shūzaburō but changed his name to Kyōsai, which is a combination of two Japanese

***Crow on a Branch.*** **ca. 1887. Kawanabe Kyōsai. Japan. Album leaf; ink and color on silk, 14.25 × 10.75 inches. Because silk has unique qualities of absorbency, the technique of painting ink on silk requires the artist to have considerable control over his materials. Kyōsai was able to achieve many grades of line sharpness in this work, thus rendering the crow as a sharp figure against a soft background (Charles Stewart Smith Collection, Gift of Mrs. Charles Stewart Smith, Charles Stewart Smith Jr., and Howard Caswell Smith, in memory of Charles Stewart Smith, 1914. Metropolitan Museum of Art).**

characters *kyō,* meaning "crazy or parody" and *sai,* meaning "studio"[15]; in other words, he embodied a kind of crazy studio art practice. Nevertheless, Kawanabe was an expert in his medium and was appreciated in the West for his individualistic manner of expression. Collectors in Europe chased after his work, particularly his crow paintings.

## Let Us Do Justice to the Raven

Pliny the Elder, Italy.

Let us do justice, also, to the raven, whose merits have been attested not only by the sentiments of the Roman people, but by the strong expression, also, of their indignation.

In the reign of Tiberius, one of a brood of ravens that had bred on the top of the temple of Castor, happened to fly into a shoemaker's shop that stood opposite: upon which, from a feeling of religious veneration, it was looked upon as doubly recommended by the owner of the place.

The bird, having been taught to speak at an early age, used every morning to fly to the Rostra, which look towards the Forum; here, addressing each by his name, it would salute Tiberius, and then the Cæsars Germanicus and Drusus, after which it would proceed to greet the Roman populace as they passed, and then return to the shop: for several years it was remarkable for the constancy of its attendance.

The owner of another shoemaker's shop in the neighborhood, in a sudden fit of anger killed the bird, enraged, as he would have had it appear, because with its ordure it had soiled some shoes of his. Upon this, there was such rage manifested by the multitude, that he was at once driven from that part of the city, and soon after put to death.

The funeral, too, of the bird was celebrated with almost endless obsequies; the body was placed upon a litter carried upon the shoulders of two Æthiopians, preceded by a piper, and borne to the pile with garlands of every size and description. The pile was erected on the right-hand side of the Appian Way, at the second milestone from the City, in the field generally known as the "field of Rediculus."

Thus did the rare talent of a bird appear a sufficient ground to the Roman people for honoring it with funeral obsequies, as well as for inflicting punishment on a Roman citizen; and that, too, in a city in which no such crowds had ever escorted the funeral of any one out of the whole number of its distinguished men, and where no one had been found to avenge the death of Scipio Æmilianus, the man who had destroyed Carthage and Numantia. This event happened in the consulship of M. Servilius and Caius Cestius, on the fifth day before the calends of April.

First century CE, Rome: a time of imperial intrigue, mutiny, and murder amidst streets filled with a million people hailing from an empire that stretched through Europe, North Africa, and the Middle East. It was an exciting age of prosperity. A new Latin phrase *imperium sine fine,* "empire without end," expressed the understanding that neither time nor space could constrain Roman reach. Indeed, one in every four people on earth lived and died under Roman rule.

Similarly, the acquisition of knowledge began to stretch in all directions. Books, as we know them today, emerged in this period. They were painstakingly handwritten onto papyrus or vellum sheets by hired scribes, then folded and bound with thread, and amassed in personal and Imperial libraries. In-house scholars were retained to collect, record, and study everything around them.

Pliny the Elder (ca. 23–79 CE) developed his skills during this time. He was an indefatigable scholar (known to work even from his bath) who is credited with creating the world's first encyclopedia, *Naturalis Historia*, a massive thirty-seven volume treatise about everything from astronomy, mathematics, and geography to the natural history of fish, plants, animals, and humankind. He wrote to Emperor Titus: "The nature of things, and life as it actually exists, are described in them [the volumes]; and often the lowest department of it. ... my road is not a beaten track, nor one which the mind is much disposed to travel over. There is no one among us who has ever attempted it,...."[16]

In *Natural History*, we find the story of a raven beloved by the people for its ability to greet members of the Forum as well as the general population. Pliny the Elder's description gives us a sense of the busy street life, the characters who passed by, and the elaborations of the bird's funeral. Events surrounding the bird's death, interment, and the execution of the murdering shoemaker took place on March 28, 36 CE.

Pliny the Elder died in the eruption of Vesuvius in 79 CE.

## Conference of the Ravens

Iceland. Excerpts from an essay by George E.J. Powell and Eiríkr Magnússon in Jón Árnason's *Legends of Iceland.*

These birds hold, every year, two great meetings, one in spring, the other in autumn. These meetings are held in every parish, and thereunto come all the ravens therein; this is, at least, sure as regards the autumn assembly. At these meetings, matters of general importance are discussed. Folk say that at their spring assembly these birds discuss plans for their subsistence during the summer; but at their autumnal meeting they chiefly discuss how they shall pair off to every farm in Iceland, for the winter. If it so happen that, at the end of the meeting, one odd raven be left, all the rest fall upon him and slay him *unguibus et rostro* [with beak and talon]. Such are the poor-law provisions of the Icelandic ravens.

## Raven Language

Raven language is a richer one than that of most Icelandic birds. In order to understand the language of ravens, it is necessary to take a living raven, and, cutting him open, to take out his heart. If the bird can move or fly two feet distance after this operation, the possessor of its

> heart will be able to understand raven language. He must keep the heart under his tongue while he listens to the black speaker, but must, at other times, keep it in a box which has never before held anything. Although everybody is not able to understand the language of this bird, everybody can at least mark the different signs in its attitude and behavior.

Ravens feature prominently in Nordic cultures. For the Vikings, few animals were more important. Raven symbols were applied to their shields and banners. Helmets and fleshforks were embellished with the bird's figure. The Viking god Odin depended upon information brought to him from two huge ravens, Hugin and Munin. They were also associated with war and death, which we explore in Chapter 5, but the ability to communicate with the supreme god Odin made them essential. Iceland itself was said to have been settled by *Hrafna-Flóki*, Raven-Flóki, who brought three ravens with him on his ship. As he neared Iceland, he released the birds to help him navigate. The first raven flew toward the Faroe Islands; the second flew back to the ship. The last raven showed Flóki the way to Iceland. Icelanders still have a soft spot for the common raven. Many believe ravens can predict weather occurrences and natural disasters and they pay attention when they hear a croak or *quork*. If only we could understand them fully....

In Icelandic folk tales, priests are oftentimes the ones able to speak the language of the ravens. One story mentions bishop Sveinn the Sapient who held conversations with ravens. Unfortunately, he developed a bad reputation because people believed he was speaking with evil raven-shaped spirits.[17] Another priest was said to have written a guide to the language and ways of the raven. And in even older times, ordinary people were sometimes able to understand raven language without the need for magic spells.

The Icelandic world of myth and story was opened to us through the efforts of Jón Árnason (1819–1888), Iceland's first librarian as well as its first curator of the Icelandic Antiquities Collection. He gathered and organized a vast array of legends. Iceland's mother tongue, known for its immense number of words used in imaginative combinations, was perfectly suited to reflect mythic encounters in a naturally dramatic landscape. Icelandic stories tend to be firmly localized in the forbidding terrain. Magical forces are strong there, as is a sense of humor. Poets could be found everywhere, and storytelling developed into a national custom. Árnason captured these tellings of dreams, visions, and appearances. He recorded some himself; others were sent to him by housewives, sheriffs, farmers, priests, even children from every nook and corner of the island. His exhaustive collection remains a great national heritage.

## Do You Understand Ravenish?

Denmark. Excerpt from "The Snow Queen" by Hans Christian Andersen.

PART THE FOURTH
THE PRINCE AND THE PRINCESS

GERDA was again obliged to stop and take rest. Suddenly a large raven hopped upon the snow in front of her, saying, "Caw!—Caw!—Good-day!—Good-day!" He sat for some time on the withered branch of a tree just opposite, eyeing the little maiden, and wagging his head, and he now came forward to make acquaintance and to ask her whither she was going all alone. That word "alone" Gerda understood right well—she felt how sad a meaning it has. She told the raven the history of her life and fortunes, and asked if he had seen Kay.

And the raven nodded his head, half doubtfully, and said, "That is possible—possible."

"Do you think so?" exclaimed the little girl, and she kissed the raven so vehemently that it is a wonder she did not squeeze him to death.

"More moderately!—moderately!" said the raven. "I think I know. I think it may be little Kay; but he has certainly forsaken thee for the princess."

"Dwells he with a princess?" asked Gerda.

"Listen to me," said the raven, "but it is so difficult to speak your language! Do you understand Ravenish? If so, I can tell you much better."

"No! I have never learned Ravenish," said Gerda, "but my grandmother knew it, and Pye-language also. Oh, how I wish I had learned it!"

"Never mind," said the raven, "I will relate my story in the best manner I can, though bad will be the best"; and he told all he knew.

Children's fairy tales deliver us to situations and conditions in which it is easy to communicate with the animals. A fairy tale frees us from Western logic and science. It is fantasy; so why wouldn't a raven know what's going on? Hans Christian Andersen (1805–1875) created many beloved stories, including "The Princess and the Pea," "The Emperor's New Clothes," "The Ugly Duckling," and "The Little Mermaid." In European literature, children's fairy tales were composed to convey moral lessons—about the nature of good and evil, the need for resilience in the face of adversity, and the benefits of love and kindness.

"The Snow Queen" was written in 1844 to tell the tale of a girl, Gerda, who does not give up until she finds and rescues her bewitched best friend Kay in the palace of the Snow Queen. Gerda's perseverance and love melts the frozen heart of Kay who quickly resumes his true form as a healthy little boy. When they return home, the children find themselves internally changed and more grown up.

Andersen drew upon familiar Nordic mythology about ravens' communicative powers to feature them talking to Gerda. A raven of the forest tries, unsuccessfully, to lead Gerda to the place where Kay is held captive. His mate, a domesticated raven, also makes an effort. They mistakenly take her to the court of a prince and princess and, while this is not where Kay is found, the birds are rewarded. The princess addresses the ravens: " 'Would you like to fly away free to the woods ... or to have the appointment secured to you as Court-Ravens with the perquisites belonging to the kitchen, such as crumbs and leavings?' Both the ravens bowed low and chose the appointment at Court, for they thought of old age, and said it would be so comfortable to be well provided for in their declining years."[18]

***"Suddenly a large raven hopped upon the snow in front of her."* 1913. William Heath Robinson. England. Illustration for *Hans Andersen's Fairy Tales.* A helpful raven stops to speak with the distraught Gerda, who is looking for her friend Kay. They have trouble communicating at first, but the raven is able to lead her to the palace of a prince and princess.**

Andersen was not the only storyteller of his time to utilize a corvid species to impart life lessons. The Brothers Grimm, famous for their collections of German tales written at about the same time, made use of the birds' more frightening features. In "The Seven Ravens," seven badly behaved boys are turned into ravens. In "The Three Crows," crows sit around a gallows-tree to trade unsavory tales then viciously attack some evil men sitting under it. And in "Faithful Johannes," three ravens talk with each other about the impending demise of the king and queen.

Andersen appreciated the more useful aspects of a raven; most Europeans associated corvids with darker dimensions—disease, death, and the devil.

## Cries of the Raven

Kwakiutl (Kwakwaka'wakw). Collected and translated by George Hunt; published by Franz Boas.

When it is desired that the owner of an after-birth should understand the cries of the raven, the after-birth is put down on the beach where the ravens peck at it. And when it is pecked at by the ravens, the man, when he is full grown, will understand the cries of the raven, for the people of olden times considered it important that the raven came to report about the arrival of warriors who came to make war upon the tribes. Then they would come at once and ask one who understands the raven, tumbling about and feathers fall[ing] out.

(Below) are the various cries of the raven, which I learned from an old man of the Kwakiutl when they discussed about it in a feast when I was a child, for when the ravens are crying, a man whose after-birth has been eaten by ravens is sent out. The one whose after-birth has been eaten by the raven understands this what I am talking about. There are only a few whose after-births have been eaten by the raven.

ga ga ga gai-----Warriors are coming to make an attack.

gax gax gax-----Ravens will eat the bodies of people drowned by the capsizing of canoes.

q!Edzō q!Edzō-----Hunters will bring much meat to feed the people.

gaga hä hägaē-----A chief (or someone else) died.

xagaq xagaq-----A woman is going to die.

k' !Emax k' !Emaq-----It will be calm weather.

sōx sōx sōx-----It will be calm and sunshine.

gūs gūs gūs-----There will be heavy rains.

wax wax wax-----A stranger will arrive on a visit.

xwo xwo xwo-----There will be a poor salmon run.

x'ok$^{u}$ x'ok$^{u}$–-----When ravens cry thus while fighting in the air, there will be bad news.

Imagine an environment teeming with sound—the water rush of salmon-filled river eddies, the tap-tapping of bird beak on deep forest tree limbs, the lift and thud of whales at sea, the croak of toads along stream banks, the deep hoot of an owl. For the Kwakiutl (now often identified as Kwakwaka'wakw) of the Northwest Coast, the world was filled with echoing life. Naturally, these sounds were important for the preservation of the birds' lives. Each thud, creak, thwap, or chirp indicated a food source or a warning. The language of the raven was no less important to the people, who were keenly sensitive and deeply knowledgeable about their surroundings. Ravens roamed the entire landscape as representatives of a clan progenitor, a culture hero, and a deity. So they were capable of divulging a lot of critical

**George Hunt (second from left, back row) with his family and Franz Boas (right). 1894. The Kwakwaka'wakw have lived for centuries along the northern side of Vancouver Island and the mainland coast across from it. It is said that their ancestors arrived in the form of animals who, when they came to a spot, shed their animal appearances, and transformed into humans. The term *Kwakiutl* was created by anthropologist Franz Boas (Courtesy of the American Philosophical Society Library).**

information about a lot of things. The raven was among the most important animals in Kwakiutl society, featuring in splendid myths, masks, and ceremonies.

Here we get practical information. We are able to study raven vocabulary. And we learn that the gift of truly understanding the cries of a raven was reserved for special people. It had to be planned for in advance—at birth. Afterward, one would be called upon to give tribal guidance. This piece was included in a document of over seven hundred pages through which German American ethnographer Franz Boas (1858–1942) presented the words of the Kwakiutl people living near Fort Rupert, British Columbia. While Boas took primary credit, the material was in fact collected by George Hunt (1854–1933), the son of a Hudson Bay Company employee and a Tlingit noblewoman who was raised in the Kwakiutl community. As a native speaker, Hunt played a crucial role in Boas's work. Hunt recorded the oral history and lore of the people, searched out and purchased artifacts for collections Boas was making, and educated Boas about the meaning of sacred objects and ceremonies. They worked together for more than forty-five years and, while Boas is today viewed as the "father of modern anthropology," Hunt's contribution to the actual material, upon which Boas' reputation is based, cannot be underestimated.

## *Vāregan:* The Voice of Heaven

Mediterranean. Analysis of a Mithraic tauroctony (bull-slaying theme) by Leroy Campbell.

The solar rays represent the Visibility Power which reveals the Way of Religion, while the black Raven whose voice orders the death of the bull represents the Word Power whose ordering utterance is Religion-in-motion, and whose flight is that of the creative wind which is warm and moist. The blackness of the raven (*vāregan* or *vāraghna*) could suggest the invisibility of the winged word that initiates Genesis in the world cave. The raven holds the edge of Mithra's mantle in his beak to indicate that his utterance is the voice of heaven represented by the sky mantle.

**Mithraic relief. 2nd century CE. Rome, Italy. Sculpted marble 38 × 55.5 × 6 inches. The "mysteries of Mithras" existed as a practice in the Greco-Roman world for about four hundred years. The cult was all male, with seven degrees of initiation (the first degree being known as "translating the Raven"). It was favored predominantly by the Roman military. The raven is found in the upper left corner. (Courtesy of the Royal Ontario Museum, Toronto, Canada. ©ROM).**

Mithraism was an ancient mystery cult lost now to obscurity. It flourished in the Mediterranean world from the first century BCE to the fourth century CE. An absence of texts confounds the attempts of scholars to explain Mithraic beliefs and practices, but Mithraic art suggests a robust and well-developed cosmogony. A central event appears to have been the slaying of a bull by Mithras. This moment is portrayed frequently in cave temples and some scholars, such as Leroy Campbell, believe the tauroctony to be a pictorial representation of ancient Persian religious ideas. Others argue that it represents a map of the stars and constellations—the raven being the constellation, Corvus. Still others believe the imagery recounts actual imperial events that took place in the Roman world. Among the many theories, Leroy Campbell's interpretation is certainly the most poetic.

According to Persian mythology, the sun god sent his messenger, the raven, to Mithras and ordered him to sacrifice the primeval white bull. At its death, the bull became the moon, and Mithras's cloak became the sky, stars, and planets. From the bull also came the first ears of grain and all the other creatures on earth. The Zoroastrian religion identifies Mithras as a god of light, truth, and the promised word; he is later thought of as the Sun god.

In this Roman rendition, Mithras is wearing the Persian costume of folded cap, tunic, and trousers. He looks directly at us at the moment of sacrifice. The raven is positioned to the far left, just below a head crowned with sun beams (the sun god). The raven is preceding the solar rays into the darkness of the cave to deliver the initiating command that ushers the universe into motion. Campbell writes: "The Word [embodied and conferred by the raven] was the moving, ruling, and more or less rational power of heaven, whereas the living bull may have represented the word incarnate, or the power of the Word that has become manifest in the flesh."[19]

## Crow Goes Hunting

Ted Hughes, England.

Crow
Decided to try words.
He imagined some words for the job, a lovely pack—
Clear-eyed, resounding, well-trained,
With strong teeth.
You could not find a better bred lot.
He pointed out the hare and away went the words
Resounding.
Crow was Crow without fail, but what is a hare?
It converted itself to a concrete bunker.
The words circled protesting, resounding.

Crow turned the words into bombs—they blasted the bunker.
The bits of bunker flew up—a flock of starlings.
Crow turned the words into shotguns, they shot down the starlings.
The falling starlings turned to a cloudburst.
Crow turned the words into a reservoir, collecting the water.
The water turned into an earthquake, swallowing the reservoir.
The earthquake turned into a hare and leaped for the hill
Having eaten Crow's words.
Crow gazed after the bounding hare
Speechless with admiration.

In an interview, poet Ted Hughes once stated: "The first idea of Crow was really an idea of style … the idea was originally just to write his songs, the songs that a Crow would sing. In other words, songs with no music whatsoever, in a super-simple and a super-ugly language which would in a way shed everything except just what he wanted to say without any other consideration and that's the basis of the style of the whole thing."[20]

In "Crow Goes Hunting," Crow tries to use human words—as weapons—but his words are swallowed up, and he is left speechless literally and figuratively. The poem is an absurdist fable about the battle between brutality (of the crow) and wit (of the hare), and wit wins as one thing repeatedly turns into another thing to foil Crow's intentions. In the middle of the poem, a liturgical repetition of the words "Crow" and "turned" creates a marvelous sense of movement that is suddenly stopped short with the last line.

The genesis of Hughes's celebrated book *Crow* was discussed in Chapter 2. What began as a Crow project in 1965 grew into a folk-epic by 1969. In this poem we see a reference to war, a topic very close to Hughes, who lived through World War II as a child and whose father suffered devastating effects from his service in World War I. Hughes uses war's wreckage for comic effect, creating a war of words and, in the process, neutralizing an experience that cannot ever be sufficiently captured in words.

# 5

# The Messenger

They materialize out of the air to colonize the tops of trees and chatter among themselves. Trading perches, first on a high branch then plonking down to a lower one, or shifting outward to weigh down a young stem, my band of three has enticed several other crows to report in about what's been going on, what they've seen and experienced. It's springtime. Chipmunks are scurrying, frogs and robins are hopping, Mayflies are swarming. The crows have checked out the territory and returned with information. If I listen hard enough, maybe I can learn a little something from them.

## *Field Notes: Roosts*

An attentive hunter, human or animal, might also want to pay attention to the activities of a crow or raven. Many native peoples know the wisdom of following a raven pack, because they regularly lead to quarry. Researcher R.D. Lawrence writes: "I have observed a number of times, ravens appear to act as scouts for wolves by congregating near prey animals, calling loudly and excitedly."[1] Wolves will follow the birds as they dart from tree to tree toward the potential of a good meal. The reverse is true as well. Wolves will alert scavenger birds to the remains of a kill with a chorus of melancholy howls. The raven-wolf relationship is one of mutual benefit; they share with one another data that is critical to their survival. The bond may go beyond even that. Ravens have been known to adopt wolves, interacting with them as they grow and following them if they leave the pack.

As we already know, crows and ravens share highly complex and detailed communiqués within their own family and friend groups. Announcements, warnings, and disclosures are often conveyed on the wing, but the roost, their overnight gathering spot at a strategically high location, serves as "central command"—an information and recruitment platform for the birds. About crow roosts, Audubon Society's Carolee Caffrey noted: "From a distance, a roost that appears to have quieted for the night is actually alive with a symphony of low-volume but highly varied crow calls, including mewing, purring, and other chatter that sounds nothing like the crow caws that we know."[2] Not sure what corvids could be conversing about, scientist Bernd Heinrich set up a raven experiment to find out. He released one set of ravens—who had been isolated—into an established roost. Another set remained isolated. The next morning when all the birds took off, Heinrich's roosting birds headed straight for a carcass he had earlier placed as bait. The birds who had remained isolated failed to show up at the bait. Clearly, the birds in the roost had informed Heinrich's visiting ravens about the location of the feast. Dr. Heinrich conducted various other experiments which proved that the birds not only briefed each other on food sources but also actively recruited one another to participate in feeding.[3] Young ravens in particular seem to operate as itinerant food-finders who then share their reconnaissance. After all, a strength in numbers gives all a better chance to get at the goodies. Heinrich sees roosts as highly fluid community centers (not all members are related) in which birds benefit from their interactions. Douglas Robinson, who conducted PhD research on crow social behavior near Newburgh, New York, concurs. He says: "We think there's some sort of cultural transmission of information taking place" at the roost.[4]

Used year after year, a roost swells at nightfall. Many people have witnessed a stream, or flight line, of birds as they proceed at dusk toward a staging area. They will travel long distances to congregate, pause, call, croak, and tussle with one another. The noise level here can be tremendous, but then suddenly the birds take off for the roost and things grow quiet. Crow roosts can get particularly large during the winter, swelling to thousands of individuals. Danville, Illinois, has been home to the largest winter roost in North America where an estimated 168,000 crows once surged in endless waves toward hickory and oak trees around Lake Vermilion. Roosts diminish during the mating season when breeding pairs separate off. Scientists still speculate about the reasons for such large crow roosts—protection from predators and the elements, nearness to food sources (crow roosts are found near reliable locations like fields and landfills), the intelligence gathering that goes on among members—these are all possibilities.

Because they are voluble, wide-ranging, and sociable (as evidenced at the roost), the corvid species seems an ideal candidate for the role of sending and receiving our own messages. Shamans of the Arctic have long seen the raven as a universal mediator, and we will explore that connection in Chapter 11. In Icelandic and Germanic folklore, the raven is known for its fortune telling powers. And both the crow and raven are identified around the world as being messengers to kings. The stories, songs, and accounts that follow reveal additional ways in which the birds carry dispatches back and forth. The Kwakiutl interpretation of raven cries as described in Chapter 4 finds an eerie echo in the ancient Tibetan manuscript that outlines how a Buddhist priest handles the utterances of the bird of Heaven. Practical prognostications can be found in India when, during the Shraddha ritual at the festival of Pitru Paksha, the mere presence of a crow pecking at offerings is a signal from the ancestors that all is well. For the Indigenous peoples of the American Plains, communities once sang and danced messages to deceased kin which were carried by the crow to the heavens. More down-to-earth information was provided by a raven to both Gilgamesh and Noah in two flood stories, written more than 2,000 years apart. Sometimes the crow delivers a deadly justice—a Korean and Italian story tell of such instances. But in at least one case the crow simply disappoints. In Japan, he simply fails at his important courier mission. Corvids have been enlisted for many tasks as messengers. In Edmund Spenser's 1590 poem "The Fairie Queene," they are the "hateful messengers of heavy things, Of death and dolour, telling sad tidings."[5] Most of the time they do their jobs well.

## Preface to the Table of Divination 

Tibet. Translated from a Sanskrit manuscript into Tibetan by Danaçila in the ninth century CE; translated from Tibetan by Berthold Laufer.

The Raven is the protector of men,
And the officiating priest carries out the order of the gods.
Sending him, the Raven into the middle of the country,
Where he has occasion for feeding on yak-flesh in the outlying pasture-lands,
The Venerable of the Gods conveys his will by means of the sound-language of the Raven.
When in the eight quarters, making nine with the addition of the zenith,
He (the Raven) sounds his notes, the three means to be observed are explained as follows:
The offering must be presented to the bird (the Raven),
And it should be a complete feeding in each instance.
In this manner, the offering is given into the hands of the god (or gods).
As to the omens, they are not drawn from the mere cries of the Raven,
But in the announcement of the omens a distinction is made between good and evil cries.
The officiating priest is in possession of the knowledge of the gods.
He teaches the orders of the gods, and it is the bird who is his helpmate in this task.
The remedies for warding off the demons are announced by the helpmate.
Truthful in his speech, he proves trustworthy,
For the Raven is the bird of Heaven;
He is possessed of six wings and six pinions.
Thanks to his visits in the land of the gods,
His sense of sight is keen, and his hearing is sharp.
Hence he is able to teach mankind the directions of the gods.
There is for man but one method of examining the sounds of the Raven,
And may you hence have faith and confidence in his auguries!
In the eight quarters, making nine with the addition of the zenith, the following sounds of the Raven occur:

> The sound *Lhou Lhou* foretells a lucky omen.
> The sound *t'ag t'ag* forebodes an omen of middle quality.
> The sound *krag krag* foretells the coming of a person from a distance.
> The sound *krog krog* announces the arrival of a friend.*
> The sound *iu ,iu* is an augury of any future event as indicated in the Table.

A mysterious roll of paper containing an extensive Table of Divination was discovered by Paul Pelliot in 1908 among dozens of religious manuscripts, paintings, and textiles that rested in the Caves of the Thousand Buddhas (located at a crossroads along a Silk Route in China's Gansu Province). Written in the ninth century CE, the Table outlines every omen delivered by the sound of a raven at every time of day in every direction depending on what the raven is doing. It further states that the wishes of a person will be fulfilled if the wish is uttered loudly at the moment the bird "speaks." It was believed by scholar Berthold Laufer that the Table derived from an Indian Sanskrit manuscript from an even earlier time as part of a broad divination practice found throughout Southeast Asia and Europe. He suggested that it was translated into Tibetan by the Mahāpandita Dānaçīla in the monastery T'an-po-c'e of Yar-kluns in the province of dBus, Central China.

From the Preface to the Table, we learn about the significance, power, and influence of the Raven who seems to connect the gods, the officiating priest, and the ordinary human. Laufer wrote: "The most significant feature revealed by this Preface, ..., is the Raven's function as the messenger of a god, so that his predictions appear as the expression of divine will. The Raven as heavenly messenger is conscious of his presages."[6] The bird translates the will of the gods; the priest, in turn, translates the Raven's utterances. They work in tandem. And the document firmly assures us that we can trust the messages of this otherwise alien creature.

Berthold Laufer (1874–1934) was a German anthropologist who, for about 35 years, was the lone sinologist working in the United States. He was recruited by Franz Boas to conduct fieldwork in China as part of the Jesup North Pacific Expedition and later led the Jacob H. Schiff expedition to China (1901–1904) where he focused on collecting everyday objects, crafts, medicinal practices, folk music, and religion. He returned having assembled a collection of over 7,500 objects, approximately 143 photographs, and 400 phonograph cylinders (the earliest known sound recordings in China). The publications that followed were numerous, ranging from monographs on material culture such as jade, amber, and pottery to the history of trade and the spread of language and ideas.

---

* The arrival of a friend should be interpreted in the figurative sense of a Buddhist blessing.

## The Kunbis

India. Commentary by Sir George Frazer.

The Kunbis, a great agricultural caste of the Maratha country in India, are firm believers in the action of ghosts, and never omit the attentions due to the ancestral spirits. On the appointed day the Kunbi calls on the crows, who represent the spirits of ancestors, to come and eat the food which he sets out for them; and if no crow appears, he is disturbed at the thought of having incurred the displeasure of the dead. So he changes the food and goes on calling till a crow comes and eats the food. From this the Kunbi infers that the first food he offered was not to the taste of his ancestors; hence, taking the lesson to heart, he continues to offer the other food to the spirits so long as a crow responds to his first invitation to partake of it. The reason why crows are taken to represent the spirits of the dead is probably connected with the widespread notion of the crow's longevity. The Hindoos believe that a crow lives a thousand years, and others think that it never dies except by violence.

**Offering to the crow during Pitru Paksha. Every year at the Indian festival of Pitru Paksha, ancestors descend in the form of crows to reside among their descendants. If a crow partakes of the food offered at this time, it means that the crow (ancestor) is pleased and will bestow blessings and good fortune in the year to come (Courtesy of Jay Mahakaal Centre of Occult Sciences).**

Ambivalence defines the general attitude in India regarding the crow. Hinduism's inauspicious and ugly widow/grandmother goddess Dhumaviti rides upon a crow. Jayant, a son of Lord Indra, was believed to have taken a crow's form in order to woo the goddess Sita. When his marriage proposal to her was rejected, he pecked fiercely at Sita and was subsequently shot in the eye by Lord Rama. Yama, god of death and justice, claims the crow as a symbol, and Lord Shani (the planet Saturn and the god of bad luck) uses the crow as his vehicle. Yet the ritual described by Sir James George Frazer in 1933 is still practiced today throughout India and involves a much deeper connection to the crow than is described by Frazer.

The ritual of *Shradh* is performed during the lunar-based sixteen-day holy period of *Pitru Paksha* when a family honors its ancestors who come down to Earth to dispense their blessings. The crow takes on a special role at this time for it is believed to embody both the form of an ancestor and that of the god Yama. At Shradh, if the food offerings laid before the crows are eaten, the ancestors are pleased. They impart good fortune on their descendants and at the same time attain for themselves spiritual energy and ultimate peace.

A website explains the event further:

> Yama vibrations are present in the departed ancestors and the same Yama vibrations are innately present in the crow. The crow is also Raja-Tama predominant [qualities of self-centeredness and greed] and just like the departed ancestors who need help. The black colour of the crow is indicative of its Raja-Tama nature. The crow has the subtle ability to see/sense departed ancestral subtle bodies. Due to their matching vibrations, a departed ancestor can enter a crow and use its body to physically partake of the food and water, thus gaining spiritual energy from it.... The crow pecking at the Naivedya [offering] indicates that the subtle body of the departed ancestor is satisfied at both levels—at the physical level by eating the food in the Naivedya through the medium of the crow and at the subtle level by imbibing subtle gases emitting from the food. This provides the energy to subtle bodies of the departed ancestors to carry on in their further journeys in the afterlife. Thus, the crow during the Shraddha ritual is a medium between the subtle bodies of the departed ancestors with unfulfilled desires and the human beings (descendants).[7]

During the emotionally and spiritually weighty time of Pitru Paksha, the crow carries blessings two ways. It strengthens both the living and the dead, and connects them in a visible, material way. Sir George Frazer reported on the practice but did not dig deep enough to understand the true significance of this important ritual. We are reminded: "The crows pecking at the *Naivedya* are tangible evidence to the fact that our departed ancestors have been benefited and that we have contributed to their well-being in the afterlife. A crow coming to peck at the *Naivedya* is the closest we would tangibly get to a departed ancestor in the afterlife. Any crow coming to peck at the *Naivedya* could well represent one of our dear departed ancestors and we get satisfaction in the knowledge that they are being helped."[8]

## Ghost Dance Songs

Arapaho (Hinono-eino), Western USA. Collected and translated by James Mooney.

### NÛ'NANÛ'NAA'TANI'NA HU'HU'-I

Nû'nanû'naa'tani'na hu'hu,'
Nû'nanû'naa'tani'na hu'hu.'
Da'chi'nathi'na hu'hu,'
Da'chi'nathi'na hu'hu.

*Translation:*
The crow is circling above me,
The crow is circling above me,
The crow having come for me,
The crow having come for me.

### NI'NINI'TUBI'NA HU'HU'–I

Ni'nini'tubi'na hu'hu,'
Ni'nini'tubi'na hu'hu.'
Nana'thina'ni hu'hu,
Nana'thina'ni hu'hu.
Ni'nita'naû, Ni'nita'naû.

*Translation:*
The crow has called me,
The crow has called me.
When the crow came for me,
When the crow came for me,
I heard him, I heard him.

### A-HU'HU' HA'GENI'STI'TI BA'HU

A-hu'hu ha'geni'sti'ti ba'hu,
Ha'geni'sti'ti ba'hu.
Hä'nisti'ti,
Hä'nisti'ti.
Hi'nisa'na,
Hi'nisa'na-
Ne'a-i'qaha'ti,
Ne'a-i'qaha'ti.

*Translation:*
The crow is making a road,
He is making a road;
He has finished it,

He has finished it,
His children,
children-
Then he collected them,
Then he collected them (i.e., on the farther side).

**BÄ'HINÄ'NINA'TÄ NI'TABÄ'NA**

Bä'hinä'nina'tä ni'tabä'na
Bä'hinä'nina'tä ni'tabä'na.
Nänä'nina hu'hu,
Nänä'nina hu'hu.

*Translation:*
I hear everything,
I hear everything,
I am the crow,
I am the crow.

**MAKA' SITO'MANIYAS**

Maka' sito'maniyañ ukiye,
Oya'te uki'ye, oya'te uki'ye,
Wa'ñbali oya'te wañ hoshi'hi-ye lo,
Ate heye lo, ate heye lo,
Maka o'wañcha'ya uki'ye.
Pte kiñ ukiye, pte kiñ ukiye,
Kañghi oya'te lo, a'te he'ye lo.

*Translation:*
The whole world is coming,
A nation is coming, a nation is coming,
The Eagle has brought the message to the tribe.
The father says so, the father says so.
Over the whole earth they are coming.
The buffalo are coming, the buffalo are coming,
The Crow has brought the message to the tribe,
The father says so, the father says so.

On the hard earth of the broad prairie and under an open, starry sky, Indigenous peoples of the Great Plains gathered for a time in the late 1800s to participate in *wanagi wacipi*, the "dance of the souls departed." The sound of their thudding feet and the lyrics of their songs rose to loved ones in the galaxy, to comfort them and bring them back to a renewed land wiped clean of the white man's influence. The Ghost Dance was part of a fervent religious movement, created by 28-year-old Northern Paiute leader Wovoka in 1889, that swept through the Plains Nations for a brief time. During

**Hide painting. 1891. Yellow Nose, Ute. Deerskin, 52 × 40 inches. A Ute captive living among the southern Cheyenne in Oklahoma painted a scene on deerskin hide for James Mooney. He portrayed Cheyenne and Arapaho men and women performing the Ghost Dance. In Yellow Nose's painting, a medicine man, at right, helps a dancer (who has leaned forward with a blue handkerchief) enter the spirit world. Already entranced figures lie inside the circle with their arms outstretched. Just below the central prone figure, whose spotted shawl has fallen to the ground, stands a woman with upraised arms holding up a sacred crow. The Ghost Dance was adopted by several different tribal groups, who integrated the dance into their own religious traditions and varied its performance. For many, wearing a decorated Ghost Dance shirt rendered dancers physically invulnerable. Ghost Dance shirts were often decorated with crow symbols (Pictures Now/Alamy Stock Photo).**

the ceremonies, people would gather in a circle holding hands. They sidestepped leftward, following the course of the sun, as they sang songs and chanted about the paradise of the future. The dancing itself served to renew and reinvigorate the worn-out world and was believed to eventually reunite the living, the buffalo, and all the dead. Many fell to the ground in a trance during which they would be transported to the afterworld to see their deceased relatives in the afterlife. The visions they experienced were recounted and used in designs for ceremonial items. The crow (*ho*), as the sacred bird of the Ghost Dance, was depicted on dancers' shirts, leggings, and moccasins. Crow feathers were worn in the hair and, if one could catch and kill a crow, its stuffed body was carried in the dance. In the hide painting commissioned and collected by James Mooney, we see a Ghost Dance in progress.

As an avatar for the Father, the crow was the messenger and leader of the spirits. In the songs, we witness the crow calling to the people, both living and deceased. He knows everything and hears their pleas. He makes a road, collects those on the other side, and advances at the head of the multitude to lead them out of the shadowland.

The Ghost Dance religion was a declaration of agency for a people who were embattled on all sides, whose lives and livelihoods were in the process of being decimated. It was a demonstration of hope and a way to deal with what was happening to them. Unfortunately, the religion was completely misunderstood by the U.S. government authorities who saw in it a great threat. They moved swiftly to quash the movement, which disintegrated after the massacre at Wounded Knee, South Dakota, on December 29, 1890, when 153 individuals, mostly women and children, were killed.

James Mooney (1861–1921) recorded our most complete account of the Ghost Dance religion. A small, energetic man with long brown hair and gray eyes, Mooney was eager to participate in collecting the so-called remains of traditional life among the Indigenous peoples of North and South America. He was sympathetic to the peoples he visited, treating their language, songs, and dances as important cultural legacies. He urged Washington bureaucrats and government soldiers to leave the Ghost Dancers alone. Mooney was working within a theoretical framework that took for granted a natural and inevitable process of "civilizing." He assumed that his job was to record and understand the "real" Indians before they changed, so that they could be helped to "civilize" themselves more effectively. Failing to understand that all cultures change, Mooney and his cohorts presented their findings as if the peoples they encountered were more genuine than those who would follow in the future. Ethnographers at that time could not imagine a scenario in which native peoples would have political agency over their own lives. Scholar Michael Elliott writes: "On the one hand, Mooney wants us to see the practices as no less valuable than their 'civilized' counterparts; on the other, Mooney does not want us to judge the Ghost-Dance religion as a set of possibly valid beliefs."[9]

## Raven and the Flood

The Epic of Gilgamesh, Ancient Babylonia. The Bible's Old Testament Genesis 7.

### The Epic of Gilgamesh

Six days and nights
The wind blew, and storm and tempest overwhelmed the country.
When the seventh day drew nigh the tempest, the storm, the battle
which they had waged like a great host began to moderate.
The sea quieted down; hurricane and storm ceased.
I looked out upon the sea and raised loud my voice,
But all mankind had turned back into clay.
Like the surrounding field had become the bed of the rivers.
I opened the air-hole and light fell upon my cheek.

Dumfounded I sank backward, and sat weeping, while over my cheek flowed the tears.
I looked in every direction, and behold, all was sea.
Now, after twelve (days?) there rose (out of the water) a strip of land.
To Mount Nisir the ship drifted.
On Mount Nisir the boat stuck fast and it did not slip away.
The first day, the second day, Mount Nisir held the ship fast, and did not let it slip away. The third day, the fourth day, Mount Nisir held the ship fast, and did not let it slip away.
The fifth day, the sixth day, Mount Nisir held the ship fast, and did not let it slip away.
When the seventh day drew nigh
I sent out a dove, and let her go.
The dove flew hither and thither,
but as there was no resting-place for her, she returned.
Then I sent out a swallow, and let her go.
The swallow flew hither and thither,
but as there was no resting-place for her she also returned.
Then I sent out a raven, and let her go.
The raven flew away and saw the abatement of the waters.
She settled down to feed, went away, and returned no more.
Then I let everything go out unto the four winds, and I offered a sacrifice.
I poured out a libation upon the peak of the mountain.
I placed the censers seven and seven,
and poured into them calamus, cedar-wood, and sweet incense.
The gods smelt the savour;
yea, the gods smelt the sweet savour;
the gods gathered like flies around the sacrificer.

### The Old Testament Genesis 7:17–8:17

Now the flood was on the earth forty days. The waters increased and lifted up the ark, and it rose high above the earth.

The waters prevailed and greatly increased on the earth, and the ark moved about on the surface of the waters.

And the waters prevailed exceedingly on the earth, and all the high hills under the whole heaven were covered.

The waters prevailed fifteen cubits upward, and the mountains were covered.

And all flesh died that moved on the earth: birds and cattle and beasts and every creeping thing that creeps on the earth, and every man.

All in whose nostrils was the breath of the spirit of life, all that was on the dry land, died.

So He destroyed all living things which were on the face of the ground: both man and cattle, creeping thing and bird of the air. They were destroyed from the earth. Only Noah and those who were with him in the ark remained alive.

And the waters prevailed on the earth one hundred and fifty days.

Then God remembered Noah, and every living thing, and all the animals that were with him in the ark. And God made a wind to pass over the earth, and the waters subsided.

The fountains of the deep and the windows of heaven were also stopped, and the rain from heaven was restrained.

And the waters receded continually from the earth. At the end of the hundred and fifty days the waters decreased.

Then the ark rested in the seventh month, the seventeenth day of the month, on the mountains of Ararat. And the waters decreased continually until the tenth month. In the tenth month, on the first day of the month, the tops of the mountains were seen.

So it came to pass, at the end of forty days, that Noah opened the window of the ark which he had made.

Then he sent out a raven, which kept going to and fro until the waters had dried up from the earth.

He also sent out from himself a dove, to see if the waters had receded from the face of the ground.

But the dove found no resting place for the sole of her foot, and she returned into the ark to him, for the waters were on the face of the whole earth. So he put out his hand and took her, and drew her into the ark to himself.

And he waited yet another seven days, and again he sent the dove out from the ark.

Then the dove came to him in the evening, and behold, a freshly plucked olive leaf was in her mouth; and Noah knew that the waters had receded from the earth.

So he waited yet another seven days and sent out the dove, which did not return again to him anymore.

And it came to pass in the six hundred and first year, in the first month, the first day of the month, that the waters were dried up from the earth; and Noah removed the covering of the ark and looked, and indeed the surface of the ground was dry.

And in the second month, on the twenty-seventh day of the month, the earth was dried.

Then God spoke to Noah, saying,

"Go out of the ark, you and your wife, and your sons and your sons' wives with you.

Bring out with you every living thing of all flesh that is with you: birds and cattle and every creeping thing that creeps on the earth, so that they may abound on the earth, and be fruitful and multiply on the earth."

Two literary masterpieces, separated in time by more than 2,000 years, tell an almost identical story about a raven sent forth from a ship at the end of a world flood. Why is that? And why was a raven selected to search for the dry land of the new Earth? Stories about the destruction of the world by water exist everywhere. More than 200 myths have been recorded from just about every continent. There are Hindu, Buddhist, Greek, and Nordic flood tales as well as Australian, Chinese, Aztec, and Indigenous North American ones. While the details vary, we usually hear of an angry god who destroys the entire world, sometimes sparing a few living beings, so that it can be remade. Some myths feature the work of animals in setting things straight but rarely do we have birds as actors.

The earliest version of a flood story was discovered carved into a clay tablet dating to the second millennium BCE. It was part of what is known as the Gilgamesh Epic which was structured in the form of a tragedy about a man's struggle with mortality. The work is considered to be one the humanity's greatest pieces of literature. The story likely goes back to the third millennium BCE, travelling orally across the Near East and Middle East. We find the same basic outline pop up within the Old

**Noah sends out the birds. 12th–13th century. Basilica, Church of San Marco, Venice. Byzantine mosaic. The Noah's ark story occupies one small section of a sprawling, multi-paneled display of Christian history and Bible stories created in St. Mark's Basilica, Venice. The scenes cover church walls, vaults, and cupolas. They gleam in radiant, golden light. In this panel, Noah tries for a second time, with the dove, to see if there is enough land on which to disembark from his boat. The raven he sent out earlier is completely occupied with its next meal (Zev Radovan/Alamy Stock Photo).**

Testament's chapter of Genesis, believed to have been completed in the sixth century BCE. The two versions differ slightly—in the Epic of Gilgamesh, the mighty gods get frustrated with human antics and send a flood to clear them out while in the Old Testament, God is determined to punish humans for their sinful behavior. At the end of the Gilgamesh story, only one couple, Utnapishtim (who narrates our section) and his wife, survive the flood and are granted eternal life. In the ancient Hebrews' Old Testament, Noah and his family do not become immortal but because of his righteousness and blameless behavior they are saved, given dominion over the animals, and can multiply. Both stories reveal much about the theology within their cultures.

So why the raven? In the Gilgamesh Epic, Utnapishtim sends out a dove, a swallow, and a raven. Noah, on the other hand, releases a raven first and then three doves. In both stories, the raven never returns. What were the boat-bound men doing? Most commentators (there have been several) believe all the birds were entrusted with the same job—to bring back information regarding the suitability of encamping on the land. A raven, or crow, was often carried on ships (hence the term "crow's nest") and released. Because the bird does not rest on water, if it did not return there was dry land in the direction it headed, and a sailor would be wise to follow. Viking and South Asian mariners used the birds for this purpose. So the story makes sense from a practical point of view (the raven did not return because there was dry land), but early rabbis and Christian church fathers believed the raven epitomized someone who had simply failed at its job. Some identified a metaphorical purpose for the raven and argued that the raven stood for sin itself leaving the ship and never returning. The philosopher Philo (ca. 20 BCE–50 CE) held that the raven and the dove personified vice and virtue (black and white). Their natures were demonstrated by their behavior. The selfish raven was waylaid by the delight of feeding upon destroyed lives while the dove could find no peace in that and returned. Still another scholar has suggested that the raven should be viewed as an embodiment of God's spirit going out: "Noah sends out the raven so that he is not just the passive recipient of, but also an active participant in, the work of God, symbolically replicating God's action through the spirit, and so appropriating it within the created realm."[10]

We have no way of knowing the true significance of the raven in the flood story. It remains a mystery. Scientific evidence for an actual world flood is earnestly debated but appears to be extremely unlikely. More likely, the theme of a catastrophic clearing-away of evil by water was too compelling to let go of. It stands as a universal myth that now circles the world. And it brings the raven along with it.

## Shoot Arrows into the Geomungo-Case

Korea. Told in the *Samguk Yusa* (Myths and Legends of the Three Ancient Korean Kingdoms). Written by Master Iryeon; translated by David A. Mason.

In 488 CE, a Dragon year, King Soji paid a pleasure-visit to *Cheon-jeong* [the Heavenly Pavilion, on a rise east of the highly sacred *Nam-san* or South Mountain]. While eating, he noticed a crow and a mouse approaching and crying loudly. To his astonishment the mouse bowed to him and spoke, "Follow the crow to where it flies," just as the crow flew away straight towards the mountain.

The king ordered an officer to follow the crow on horseback to the tiny Yangpisa Hamlet, right at Nam-san's foot, which had a pretty pond. The horseman was distracted by two pigs fighting, and when he finally looked for the crow again it had vanished. As he walked on the pond's edge wondering what to do, an old man [the *Sanshin,* mountain-spirit, of Nam-san] arose out of the water and presented a sealed letter to him.

In amazement the officer rode back to the pavilion and gave the letter to King Soji, who noticed that on the envelope was written, "If opened, two people will die; if not opened, one person will die." The king then said, "It is better not to open the letter and let one man die, than to cause two people to die, right?" But the attendant royal astrologer responded, "The two people referred to are commoners, but the one person can only be our monarch."

The king agreed with this wise insight and opened the envelope. There was a paper letter inside, upon which was written, "Shoot arrows into the Geomungo-case in your bedroom." [Geomungo is an ancient Korean instrument like a Chinese zither but with fewer and heavier strings; several would have been kept in a large wooden box like a closet, with a cloth door.]

The king hurried back to the palace and stormed into his bedroom. He shot several arrows into the Geomungo-case and heard cries of pain. When he ripped open the case's door, there were the dead bodies of his queen and the palace's chief Buddhist monk. He realized that they had been making love in the bed and had jumped into the case to hide. He further realized that they had been planning to overthrow and murder him. He indeed would have been the one person to die instead of these two if he had not obeyed the letter. He owed great gratitude to the *Nam-san Sanshin*!

From that time forward it became customary to be cautious in words, behavior, and travel on the First Moon's pig day, rat day, and horse day, and to observe each year's First Full Moon as the "Crow's Veneration Day." On that day, cooked glutinous rice is offered on the village altars in tribute to the black bird that saved the king's life and brought deserved death to the adulterous queen and her priestly lover.

**The pond of the *Nam-San* (South Mountain) spirit. Gyeongju, South Korea. The pond to which the crow flew became known as *Seochul-ji* [Writing-Origin Pond]. A lovely Neo-Confucian wooden pavilion named *Iyo-dang* was built beside it in 1664 for contemplation and commemoration of the tale. Lotus flowers and tall grasses grow from the shallow waters, and excellent old trees line its raised banks (photograph courtesy of David A. Mason).**

From the mists of the past comes a legend about a king's appreciation for a crow. The term legend best defines this story because King Soji was indeed the 21st ruler of the kingdom of Silla, a small territory existing in the southeastern part of the Korean peninsula which grew to rule the entire area from the seventh to eleventh centuries CE. The king was real; his legendary exploits may have been fabricated.

The story reaches far back in time but lives among Koreans today. A Buddhist priest, Iryeon (1206–1289 CE), set about to record stories from his country. His *Samguk Yusa* drew upon the beliefs and folklore associated with the history of Korea during its "Three Kingdoms" period (57 BCE–668 CE). It was written in literary Chinese at a time when the land was dominated by Chinese Mongols, and served to both confirm and celebrate the uniqueness of Korean history and spirit. Iryeon's work, carved onto wooden plates, was lost to the physical ravages of time. In 1512, a nobleman named Yi Kye-pok was able to obtain one copy. He wrote: "It is natural that old things are lost and lost things are found again as nations and peoples rise and fall. Realizing this, I have had this book reprinted to preserve permanently a literary treasure of our scholars for all ages to come."[11] From this sixteenth-century document (first translated into English in 1972), we learn about the crow's role in saving King Soji.

It was a bittersweet task that the crow performed for it was this "black bird that saved the king's life and brought deserved death to the adulterous queen and her priestly lover." The crow was a silent messenger—leading the king's officer to a cryptic mountain spirit who held out the ominous sealed letter. Nevertheless, the bird was considered worthy of honor for its actions.

Today, echoes of the legend can be found in *Daeboreum*, a joyous holiday marking the first full moon of Korea's lunar year (the "Crow's Veneration Day"). Many traditional customs are practiced at this community-centered celebration—roaming bands of musicians play their instruments to the footsteps of flag-bearing dancers, daytime games of tug-of-war are followed by nighttime bonfires, and small gestures such as the cracking of nuts are thought to ward off bad luck and bring good fortune in the new year. Sadly, the crow's responsibility for this happy tradition has largely been forgotten.

## Apollo's Raven

Ovid, Italy.

Coronis of Larissa was the loveliest girl in all Thessaly.... But that bird [the Raven] of Phoebus [Apollo] discovered her adultery and, merciless informer, flew straight to his master to reveal the secret crime. The garrulous Crow followed with flapping wings, wanting to know everything, but when he heard the reason, he said "This journey will do you no good: don't ignore my prophecy! See what I was, see what I am, and search out the justice in it. Truth was my downfall."

To all this, the Raven replied "I pray any evil be on your own head. I spurn empty prophecies" and, completing the journey he had started, he told his master he had seen Coronis lying beside a Thessalian youth. The laurel fell from the lover's head on hearing of the charge, his expression and colour and the tone of his lyre changed, and his mind boiled with growing anger. He seized his usual weapons, strung his bow bending it by the tips, and, with his unerring arrow, pierced the breast that had so often been close to his own. She groaned at the wound, and as the arrow was drawn out her white limbs were drenched with scarlet blood and she cried out "Oh Phoebus it was in your power to have punished me, but to have let me give birth first: now two will die in one." She spoke, and then her life flowed out with her blood. A deathly cold stole over her body, emptied of being.

The appearance of a raven (or a crow) is not always a good thing. Sometimes it delivers bad news. This was the case for the god Apollo, son of the great god Zeus, when the raven (against the advice of the crow) flew with all speed to Apollo's side to tell him that his beloved Coronis was cheating on him with a boy from Thessaly.

**Cylix (shallow drinking cup with two handles). 480–470 BCE. Delphi, Greece. White ground terracotta, 7-inch diameter. The Greek god Apollo appears as a youthful figure. He wears a laurel wreath wrapped in his elaborate hairdo along with clothing that suggests the status of a god. He is poised on a lion-footed chair. In one hand he holds a seven-stringed lyre; in the other a bowl for pouring libations. Ritual purification by water was an important aspect of the cult of Apollo and he may be practicing this rite himself. The dark figure of a raven boldly faces the god at eye level. Perhaps the bird has just landed to tell Apollo about his beloved's betrayal (Greek Ministry of Culture).**

In some versions of the story, the raven was at that time white but burned black by Apollo in his fury; in other versions Coronis herself turns into a crow (hence the Greek word for crow is *corone*). Here, in Ovid's version, Apollo brutally pierces her body with an arrow. The story continues but it all ends badly; Apollo is deeply remorseful, and the raven is in bad shape. Thus we learn about the risks of being a messenger.

The travails of Apollo, Coronis, and the raven were widely known around the Roman Empire at the time Publius Ovidius Naso (Ovid, 43 BCE–17/18 CE) wrote his extraordinary opus *Metamorphoses*. In the work, Ovid reshaped familiar, previously published stories to suit his own poetic and philosophical purposes, which most often concerned the theme of love in all its forms. *Metamorphoses*, an elegiac epic that uses a mythic framework to chronicle the history of the world from Creation to Ovid's present day, interweaves the passions and exploits of the gods with those of humankind. The fury and consequences resulting from spurned love is dramatized in this episode. We hear from Ovid about another consequence of love in Chapter 10.

## Song of the Owl God

Ainu, Japan. Translated by Donald L. Philippi.

"Long ago,
when I used to speak,
my voice would ring out
like the buzzing
at the center of the handgrips
of bows wound with cherry
bark,
but now
I feel old,
I feel feeble.

O for someone
who is eloquent
enough to be trusted
with a message!

If only there were such a one,
I would send him as a messenger
to the heavens
bearing five
and a half messages!"

While saying these words,
I beat time
on top of the lid
of a wine-tub with a hoop
around it.

Then someone [appeared] at the door and said:
"Who but me
is eloquent enough to be
trusted with a message?"

I looked and saw
that it was
Crow Boy.
I invited him in.
Then
I beat time

on top of the lid
of the wine-tub with a hoop
around it
while I recited the message
which
Crow Boy
was to bear.

Three days went by.

While I was just reciting
the third message,
I looked up and saw that
Crow Boy
had dozed off, nodding his
head,
behind the hearth frame.

When that happened,
I flew into
a terrible rage.
I thrashed Crow Boy,
feathers and all,
and killed him.

Then once again,
I began to beat time
on top of the lid
of the wine-tub with the hoop around it,
....

This is the opening portion of a long song during which the elderly Owl God, living in the land of humans as their guardian, must get a message to other gods in the sky about a famine that has broken out on Earth. He sings to us a lament about the lazy Crow Boy who could not deliver the message to the heavens. He had to beat that wine-tub drum again and again. The Owl God finally succeeds with the dipper bird and is told that to save them he must teach humans proper hunting and fishing rituals. At the end of the song, his mission complete, the Owl God ascends into the heavens, leaving behind a youthful warrior to watch over the humans.

The crow's offense was to fall asleep during the (three-day-long) recitation of the message, for which he was killed.

During the performance of this Ainu song, the reciter assumed the persona of the deity and sang his words. Listeners gathered around to hear from the god himself, learning directly about his adventures from his point of view and following his unfolding understanding. Other gods were believed to be in the audience as well, cloaked in invisibility or in the form of an animal. The first-person narrative was a special characteristic of Ainu epic songs; their oral literature was fundamentally about self-revelation. In this instance, the Owl God sings his song here on Earth. In Chapter 10, we hear from the Carrion Crow goddess about her performances in the heavens.

Doing a better job than the crow, Donald Philippi (1930–1993) brings us stories from the early Ainu peoples of Japan. Philippi said of his work: "A translator is like a transmitter of messages from one world to another,…."[12] Indeed, his efforts to collect and examine Ainu epic poetry bridge a wide chasm, inviting access to a society that was deeply rich in folklore and song. Mentioned in Chapter 3, the Ainu of the Hokkaido, Sakhalin, and Kuril Islands were a pre-literate people racially and ethnically distinct from the Japanese. They lived in relative isolation until the late nineteenth century without the need for domesticated animals, rice cultivation, or iron and bronze tools. Their culture and religion were founded on a belief that *ainu* (humans), animals, and *kamui* (non-human species) could interact across boundaries for the benefit of all. Interdependence and inter-species communication were foundational concepts. And the mythic epics (*kamuiyukar*), sung with musical burdens (a chorus or refrain), provided the means for cross-species communication, for obtaining blessings, and for driving away evil. The epics could be sung at any time and were repeated often, each time slightly differently and accompanied by the murmur of listeners repeating the burdens back.

Unfortunately, the Ainu experienced the same kind of erasure that many other Indigenous peoples around the world have. The Japanese government set out to forcibly assimilate them and they live today as a small, marginalized group that struggles to maintain its culture and traditions.

# 6

# A Closeness with Death

A black figure has alighted on a grisly carcass. Then, a strange flap of wings and uneven burst of *caws* ensues. Phantoms quickly descend to poke and pull at the decomposing flesh. This image, seen everywhere, buries itself deep within our unconscious, both disturbing and fascinating us. The raven and the crow are consummate scavengers, ready to take advantage of what others have left behind. They have always been associated, literally and metaphorically, with the ultimate "otherness" of death, becoming both followers and harbingers of our departure from life. And, because we humans create so much death through war, they are also linked to aspects of the battlefield. Across the world, myths and legends offer us the details of corvid commerce with life's final struggle. But what do the birds themselves understand about death?

## *Field Notes: Awareness of Death*

Folk knowledge tells us that crows will often noisily muster around a fellow bird who has died. First-person accounts usually describe a commotion, followed by a gathering around the dead bird, then silence, then the birds are gone. Ornithologist John Marzluff writes, "it is the *silent departure* that seems so uncrowlike to witnesses."[1] Other bystanders and owners of pet crows and ravens describe demonstrations of what we would call "emotions" at the death or disappearance of friend or kin. Naturalist Lyanda L. Haupt watched as crows congregated around an injured bird and just sat there with it quietly: "This was a crow hospice, and it seemed the attendant birds were simply 'being with' the dying crow, just as we are asked by human hospice workers to be present to our own beloved dying. This was a sweet, arresting moment to witness, and I have since seen other instances of crows gathering with attentive care around sick or injured birds,...."[2] Both John Marzluff and ornithologist Kevin McGowan observed American crows caring for birds. And Charles Trost, a zoologist at Idaho State University, claims to have observed ravens holding "funerals" for their dead and then refusing to return to a kill site.[3] Trost also spent time monitoring magpie (also a corvid) "funerals." They would "assemble in a nearby tree, starting a commotion to alert the flock. Then, one by one, they fly down to stand beside the body—or the feathers—to deliver a brief oration while the others listen. They only do this once. Then, everybody leaves."[4]

Scientists resist the tendency to attribute human characteristics to bird or

**Flock of hooded crows (*Corvus corone cornix*) around the body of a dead red fox. Norway. Crows are cautious when first approaching food, and ravens will wait a long time for someone else eat first. Both birds have long experience with poisoned carcasses. Once they find it safe, a hierarchy of access is put in place. Breeding males or mated pairs get in first, followed by the others (blickwinkel/Alamy Stock Photo).**

animal behavior. While very few species pay attention to their deceased members (the list comprises chimpanzees, elephants, dolphins, and humans, as well as corvids), even fewer rigorous scientific studies have been conducted on the topic. They might reveal more mundane reasons for the bird behavior. Dr. Marzluff suggests that the birds may bunch around a dead bird in order to work out a shift in their hierarchy. Or, they may be acquiring new knowledge about the environment. He notes that their learning is facilitated by the high sensitivity in their brains' amygdala to the release of stress hormones.[5] Dr. Kaeli Swift of the University of Washington set out to explore the question. She sent a volunteer to a feeding site with a dead crow in hand. The volunteer was viciously mobbed by its crow group. Another volunteer then approached with a dead pigeon and was only sometimes mobbed. The crow elicited a much stronger "emotional" response. Swift also noted that, after seeing a volunteer with a dead crow, the birds took much longer to approach their food while, conversely, the sight of a dead pigeon had no effect on feeding behavior.[6] Her scientific reports suggest that crows and ravens pay attention because a dead bird might be a sign of danger (hence they jump around and make loud warning calls to one another). "Group members might be learning to avoid an area or others like it in the future for their own safety. Such learning could explain why crows stop frequenting areas where they have been trapped or killed."[7]

We already know the species to be great communicators with good memories. But empathy? Swift also began to investigate that idea. Her team showed a dead crow to another crow, then scanned the live crow's brain to see what section lit up. The team discovered that the decision-making area was aglow; the part of the brain where a bird determines whether there is danger and what to do about it. That left Swift with her original question: would the amygdala (emotion center) light up if the bird saw its dead mate? She doesn't yet have an answer but says it's possible. Studies of ravens have shown that they are capable of something called "emotion contagion," a bird who looks at another bird's reaction to something and behaves as if they feel the same way.[8] Perhaps her crow subjects are capable of transmitting more than factual information.

Ravens and crows certainly know something about the practical aspects of death. They are sure to show up on a battlefield to partake of flesh from all sides. In the old days, they appeared at the gallows (hence the English term "Ravenstone" for a place of execution). Today, they patrol the highways for fresh kills. The birds are scavengers and, less commonly, killers. Naturalist David Rains Wallace writes that ravens have been seen killing polar sea mammals, cattle on the range, and newborn lambs.[9] It is not unheard of for crows to kill small animals and Dr. Marzluff described a scene where thirty American crows attacked a grounded crow whose wing was broken.[10] But broader, more nuanced stories exist in world literature. Our tellings convey very complex and ambivalent feelings. Because of their closeness to death, we bequeath to corvids power, mystery, knowledge, and magic, but also our deep uneasiness.

## The Raven Teaches the Man

Jewish. Commentary of Rabbi Eliezer the Great.

And Cain rose up against Abel his brother, and slew him. Cain spake before the Holy One, blessed be He: Sovereign of all the worlds! "My sin is too great to be borne," for it has no atonement.

The dog which was guarding Abel's flock also guarded his corpse from all the beasts of the field and all the fowl of the heavens. Adam and his helpmate were sitting and weeping and mourning for him, and they did not know what to do (with Abel), for they were unaccustomed to burial. A raven (came), one of its fellow birds was dead (at its side).

(The raven) said: I will teach this man what to do. It took its fellow and dug in the earth, hid it and buried it before them.

Adam said: Like this raven will I act. He took the corpse of Abel and dug in the earth and buried it.

The Holy One, blessed be He, gave a good reward to the ravens in this world. What reward did He give them? When they bear their young

> and see that they are white they fly from them, thinking that they are the offspring of a serpent, and the Holy One, blessed be He, gives them their sustenance without lack....
>
> He giveth to the beast his food, and to the young ravens which cry.

The raven plays a role in a dramatic story that is shared by three great religions—Judaism, Christianity, and Islam. It is the story of humanity's first murder, first fratricide, and first burial. Cain and Abel, sons of Adam and Eve, competed for God's love and attention. God favored Abel; Cain then murdered him. Its gripping narrative has been told and retold through the centuries. We still see allegorical references to it in modern movies, TV series, and graphic novels. The stunning event, supposed to have occurred at the beginning of the human experience, is told in Genesis without reference to the raven.

Rabbi Eliezer (ca. 40–120 CE), or a later writer posing as the famous Rabbi (scholars are not certain), is credited with introducing the bird to the story and tying it at the end to the very ancient Psalm 147. He most likely was drawing from Judaic folklore.

The raven we read about here is an actor in the story. He desires to educate Adam and with his actions he shows Adam how to bury his son. The first burial (the instruction really an act of mercy on the part of the raven) intrinsically holds deep psychological, emotional, and spiritual weight. What could be worse than burying your son who has been killed by his own brother?

The Islamic Qur'an (seventh century CE) also recounts this moment in Surah (chapter) 5. In this telling, however, Cain buries Abel and the raven (sometimes identified as a crow) is not so much an actor as a messenger from Allah. Sura 5:31 reads: *Thereupon Allah sent forth a raven who began to scratch the earth to show him how he might cover the corpse of his brother. So seeing he cried: "Woe unto me! Was I unable even to be like this raven and find a way to cover the corpse of my brother?" Then he became full of remorse at his doing.*[11]

At the end of the Jewish episode, we learn that God rewards all ravens—in an odd sort of way. We are informed that ravens are born white and abandoned by their parents. God provides them sustenance. Rabbi Eliezer should have consulted a local ornithologist. Ravens do not abandon their young.

## Crow Flying

Nabaloi Igorot song, Philippines. Collected and translated by C.R. Moss.

Crow flying
Go north to Loo;
Go look down at
The hanging head.
Croaking crow,
What are you looking at?
Go look down at
The body we beheaded,
The body we beheaded.

The Upperworld was maintaining a residence for birds, sky deities, and spirits of the dead when Claude R. Moss encountered the Nabaloi Igorot people of Luzon, the Philippines, around 1915. Deities took a human form in the Upperworld and donned an avian form on Earth, where they would swoop and soar over all of humankind's activities. The songs and chants directed to them, such as "Crow Flying," brought appeasement and good luck.

"Crow Flying" was sung at a *bindayan*, a head-hunting celebration. When Moss studied the music and ceremonies of the Nabaloi Igorot, head-hunting had mostly disappeared as a practice. However, the bindayan remained as an important cultural ritual, given to cure sickness or in fulfillment of a vow. In the depths of the night, villagers gathered at a campsite, taking with them spears, shields, jars of *tapuy* (alcoholic drink), and rice. The *mambunong* priest would call the names of ancestors and address them as follows: "You celebrated the bindayan, which we are celebrating now. Help us to remember the song; help us to remember the names of the brave head takers."[12] Songs, ritual dances, magic formulas, whispered prayers, and the retelling of mythic headhunts were recalled and passed down through the generations in this way.

The bindayan originally celebrated the deeds of brave warriors (now gods) who took heads from other villages. The heads were placed on poles, around which the villagers danced. Head-hunting was conducted until the late nineteenth century throughout Southeast Asia and the Philippines. Westerners found the tradition shocking and it disappeared shortly after their arrival, leaving ethnographers to disagree about its purpose in the social and religious fabric of Indigenous life. Undoubtedly, head-hunting served several purposes. On a practical level, the taking of a head increased one's reputation and enabled one to attract a wife. Socially, it might be conducted for revenge. Religiously, head hunts were equated with cosmic journeys and because the soul was thought to reside in the head, the capture of a head added to the general corpus of souls in the community. But not only that.

Dr. Robert McKinley writes: "All known headhunting groups in Southeast Asia and Oceania tend to do something with the head (or with the name associated with the head) which symbolizes the incorporation of the former enemy as a new member of one's own society. The enemy becomes a friend, and often a good friend. As a vehicle for spirits, the enemy's head brings fertility and many other mystical benefits."[13]

By the twentieth century, head-hunting was no longer practiced. Carved heads replaced real ones for the important rituals that persisted to recreate a cosmic reality. The song "Crow flying" asks the crow-as-spirit to acknowledge and celebrate the head-hunting tradition.

Ethnologist Claude R. Moss (1875–1958) lived for twelve years among the Nabaloi people, a non-literate society with a complex form of polytheism and a deep connection to the land and to the generations of its ancestors. He recorded and translated Nabaloi songs which he sent to Teodoro Francisco (a school band instructor in Kabayan, the Philippines) to transcribe into a music staff for Alfred L. Kroeber, an eminent anthropologist at the time, to analyze. Kroeber, Moss, and Francisco encountered much difficulty transcribing the music for Western ears.

## Hugin and Munin, Odin's Ravens

Óláfr hvítaskáld Þóðarson and Snorri Sturluson. Scandinavia.

*Flugu hrafnar tveir af Hnikars ǫxlum;*
*Huginn til hanga, en á hræ Muninn.*

Two ravens flew from Hnikarr's [Óðinn's] shoulders;
Huginn to the hanged one, and Muninn to the corpse.

Fragment 4 of a poem by Óláfr hvítaskáld Þóðarson

Two ravens sit on Odin's shoulders, and bring to his ears all that they hear and see. Their names are Hugin and Munin. At dawn he sends them out to fly over the whole world, and they come back at breakfast time. Thus he gets information about many things, and hence he is called Rafnagud (raven-god). As is here said:

Hugin and Munin
Fly every day
Over the great earth.
I fear for Hugin
That he may not return,
Yet more am I anxious for Munin.

from the *Prose Edda*, by Snorri Sturluson

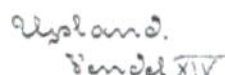

**Vendel XIV helmet and illustration. Vendel Period (550–750 CE). Sweden. The raven appeared frequently on Viking helmets. Here we see a raven drop down to become a nose protector. Engraved images on the helmet feature an army of raven-headed men who stand ready for battle. Perhaps these soldiers obtained their military prowess from the raven, the supreme god Odin's special bird (Photo: Katarina Nimmervol / Courtesy of the Swedish History Museum [CC BY]).**

Dispatched by the supreme god to roam the land and keep tabs on the ebb and flow of human battle, a Viking raven's sacred job brought it into daily contact with the human habit of carnage. Odin, the Nordic deity of war, death, magic, and poetry, kept two ravens with him at all times. Hugin ("Thought") and Munin ("Memory") reported back on who was winning a battle, what side was in trouble, and how many men were bound for the afterlife. The *valkyrie*, supernatural women warriors who usually assumed raven form, escorted those who died in battle to Valhalla, a gleaming Hall of the Fallen. So why does Odin (also known as Rafnagud, raven-god) worry that Hugin and Munin might not return? Perhaps, once on the battlefield, they might stop to nibble, linger to think, and risk being killed themselves.

The raven's association with war and death is found throughout northern Europe's mythology and iconography; most well known is the raven banner that was carried into battle by Viking chieftains during the ninth to eleventh centuries. The bird's supernatural military intelligence capabilities were prized during the Viking age in Scandinavia where their earthly presence was a portend that something important was about to happen. Ravens were thought to have knowledge of the future which they would share with those who were kind to them, and if a warrior could enter the world of ravens, he would have special abilities indeed.

Óláfr hvítaskáld Þóðarson (ca. 1210–1259) and Snorri Sturluson (1178–1241) were celebrated Icelandic skaldic poets of the Scandinavian Middle Ages who recited legends, histories, and myths to those gathered at the courts of kings and noblemen. They were also nephew and uncle. The *Prose Edda* (ca.1225) created by Snorri Sturluson is certainly the most famous piece of Icelandic literature. It contains a collection of poems that were composed in distinctive Nordic literary forms to describe the characters and exploits of pre-Christian gods and heroes.[14] Icelandic scholar Guðrún Nordal writes that "skaldic poetry was practised at the highest level of society from the ninth century to the middle of the fourteenth, … the verse served not only the ends of the secular ruling class, but it belonged at the heart of pagan religion as well as that of the Christian Church and its learned milieu."[15] The poetry entertained and educated everyone.

## The Dream of Rhonabwy

Wales, Great Britain. Translated by Lady Charlotte Guest.

[Rhonabwy (a retainer of Madog, the ruler of the Powys area of central Wales) is traveling across the countryside with companions when they come to a house. Rhonabwy falls asleep and dreams. In the middle of the dream, he encounters the mighty King Arthur of the past sitting in a great chair on a carpet:]

And Arthur sat within the carpet, and Owain the son of Urien was standing before him. "Owain," said Arthur, "wilt thou play chess?" "I will, Lord," said Owain. And the red youth brought the chess for Arthur and Owain; golden pieces and a board of silver. And they began to play.

And while they were thus, and when they were best amused with their game, behold … there came a young page.… And he came to the place where the Emperor and Owain were playing at chess.

Then said the youth unto Owain, "Lord, is it with thy leave that the young pages and attendants of the Emperor harass and torment and worry thy Ravens? And if it be not with thy leave, cause the Emperor to forbid them." "Lord," said Owain, "thou hearest what the youth says; if it seem good to thee, forbid them from my Ravens." "Play thy game," said he. Then the youth returned to the tent.

That game did they finish, and another they began, and when they were in the midst of the game, behold, a ruddy young man with auburn curling hair and large eyes, well-grown, and having his beard new-shorn, came forth from a bright yellow tent, upon the summit of which was the figure of a bright red lion. … And he came to the place where Arthur and Owain were playing at chess. And he saluted him. And Owain was troubled at his salutation, but Arthur minded it no more than before. And the youth said unto Owain, "Is it not against thy will that the attendants of the Emperor harass thy Ravens, killing some and worrying others? If against thy will it be, beseech him to forbid them." "Lord," said Owain, "forbid thy men, if it seem good to thee." "Play thy game," said the Emperor. And the youth returned to the tent.

And that game was ended and another begun. And as they were beginning the first move of the game,.... they saw a youth with thick yellow hair upon his head, fair and comely, and a scarf of blue satin upon him, and a brooch of gold in the scarf upon his right shoulder as large as a warrior's middle finger. ... In the hand of the youth was a mighty lance, speckled yellow, with a newly-sharpened head; and upon the lance a banner displayed.

Fiercely angry, and with rapid pace, came the youth to the place where Arthur was playing at chess with Owain. And they perceived that he was wroth. And thereupon he saluted Owain, and told him that his Ravens had been killed, the chief part of them, and that such of them as were not slain were so wounded and bruised that not one of them could raise its wings a single fathom above the earth. "Lord," said Owain, "forbid thy men." "Play," said he, "if it please thee." Then said Owain to the youth, "Go back, and wherever thou findest the strife at the thickest, there lift up the banner, and let come what pleases Heaven."

So the youth returned back to the place where the strife bore hardest upon the Ravens, and he lifted up the banner; and as he did so they all rose up in the air, wrathful and fierce and high of spirit, clapping their wings in the wind, and shaking off the weariness that was upon them. And recovering their energy and courage, furiously and with exultation did they, with one sweep, descend upon the heads of the men, who had erewhile caused them anger and pain and damage, and they seized some by the heads and others by the eyes, and some by the ears, and others by the arms, and carried them up into the air; and in the air there was a mighty tumult with the flapping of the wings of the triumphant Ravens, and with their croaking; and there was another mighty tumult with the groaning of the men, that were being torn and wounded, and some of whom were slain.

And Arthur and Owain marveled at the tumult as they played at chess; and, looking, they perceived a knight upon a dun-colored horse coming toward them.....He had in his hand a blue-shafted lance, but from the haft to the point it was stained crimson-red with the blood of the Ravens and their plumage.

The knight came to the place where Arthur and Owain were seated at chess. And they perceived that he was harassed and vexed and weary as he came towards them. And the youth saluted Arthur, and told him that the Ravens of Owain were slaying his young men and attendants. And Arthur looked at Owain and said, "Forbid thy Ravens." "Lord," answered Owain, "play thy game." And they played. And the knight returned back towards the strife, and the Ravens were not forbidden any more than before.

> And when they had played awhile, they heard a mighty tumult, and a wailing of men, and a croaking of Ravens, as they carried the men in their strength into the air, and, tearing them betwixt them, let them fall piecemeal to the earth. And during the tumult they saw a knight coming towards them, on a light grey horse, and the left foreleg of the horse was jet-black to the center of his hoof....
>
> And the youth saluted the Emperor: "Lord," said he, "carest thou not for the slaying of thy pages, and thy young men, and the sons of the nobles of the Island of Britain, whereby it will be difficult to defend this island from henceforward for ever?" "Owain," said Arthur, "forbid thy Ravens." "Play this game, Lord," said Owain.
>
> So they finished the game and began another; and as they were finishing that game, lo, they heard a great tumult and a clamor of armed men, and a croaking of Ravens, and a flapping of wings in the air, as they flung down the armor entire to the ground, and the men and the horses piecemeal. Then they saw coming a knight on a lofty-headed piebald horse.....
>
> Wrathfully came the knight to the place where Arthur was, and he told him that the Ravens had slain his household and the sons of the chief men of this island, and he besought him to cause Owain to forbid his Ravens. And Arthur besought Owain to forbid them. Then Arthur took the golden chessmen that were upon the board, and crushed them until they became as dust. Then Owain ordered Gwres the son of Rheged to lower his banner. So it was lowered, and all was peace.

Raven stories are prized throughout Great Britain, but Wales is not Scotland, is not Ireland, is not England. Each area's stories are unique. In Ireland, the war goddess Badb takes the form of a crow to cause terror among opposing forces, while Scotland's tempestuous old hag Cailleach Bheur, goddess of winter, appears unbidden as a raven. Cornwall's coat of arms features a black Chough (a corvid member) with red beak and red feet to represent King Arthur's last battle when his spirit entered the bird, and England's Tower of London harbors ravens because the head of *Bendigeidfran* ("Brân the Blessed," guardian of Britain whose totem is the raven), is buried beneath Tower Hill. From Wales, we have "The Dream of Rhonabwy," a dream sequence that features a band of ravens with supernatural strength.

In the extraordinarily theatrical episode found in *The Mabinogion* (only a portion of which is included here), King Arthur and Owain mab Urien sit down to a game of chess. During the game they are informed that Arthur's men are brutally attacking Owain's raven band. After much back and forth, during which Arthur repeatedly ignores reports of slaughter, Owain commands his attendant to raise aloft his standard flag. The ravens, with their magical powers, are then able to tear Arthur's men to shreds. Arthur finally crushes the golden chessmen on the chessboard, signaling peace. The ravens in this melodrama are capable of seizing "some by the heads and others by the eyes, and some by the ears, and others by the arms" to carry the soldiers into the sky and to their deaths. The birds make a perfect battle-hardened army for Owain mab Urien, who was in fact a historical figure living in the sixth century (and whose family coat of arms contains three ravens on a white field). The association of Owain with ravens has a long history in Welsh tradition and is frequently mentioned in Welsh poetry, so the scene would have delighted listeners.

*The Mabinogion* is a collection of Wales's earliest literature, written down between the eleventh and thirteenth centuries but deriving from much earlier oral literature. It stands as a cultural treasure and invaluable storehouse of Welsh language, culture, and folklore. "The Dream of Rhonabwy" is particularly interesting to scholars because it begins in an actual time-location (mid-twelfth-century Powys, central Wales) but soon describes events that occurred during the time of the legendary King Arthur (fifth–sixth century). The story's structure (a dream-vision time-travel) is intriguing because it includes a variety of symbolic content as well as magico-ritualized action. History and myth attach to one another and move together through time. The symbolic content also seems to point to a sophisticated use of irony, even parody, by the original author. Some scholars feel that it was composed to comment, unfavorably, on times past.[16] The true meaning of "The Dream of Rhonabwy" remains elusive to readers but the drama and setting are truly captivating.

Lady Charlotte Guest (1812–1895) was enchanted by *The Mabinogion*. On her way to becoming an accomplished linguist, educator, collector, philanthropist, society hostess, expert in iron production, and mother of ten children, Guest translated and published the first modern print version of *The Mabinogion*, which totaled seven volumes. She was a champion of the Welsh literary renaissance of the nineteenth century.

## The Death of Cuchulainn

Ireland. Retold by Charles Squire.

Cuchulainn drank, and bathed, and came out of the water. But he found that he could not walk; so he called to his enemies to come to him. There was a pillar-stone near; and he bound himself to it with his belt, so that he might be standing up, and not lying down. His dying horse, the Gray of Macha, came back to fight for him, and killed fifty men with his teeth and thirty with each of his hoofs. But the "hero-light" had died out of Cuchulainn's face, leaving it as pale as "a one-night's snow," and a crow came and perched on his shoulder.

Now that they were certain that Cuchulainn was dead, they all gathered round him, and Lugaid cut off his head to take it to Medb [Queen of Connacht].

*The Death of Cuchulainn*. 1935. Oliver Sheppard. General Post Office, Dublin. Bronze. For the Irish, Cuchulainn symbolizes ultimate courage in the face of death. He has become linked with cultural nationalism and the political independence movement. The crow atop his shoulder is the Morrigan, an ominous war goddess whose love he spurned (Phil Crean A/Alamy Stock Photo).

The moment of death for the superhuman culture hero of Ireland, Cuchulainn, is almost always portrayed with a crow (or raven) perched on his shoulder. In Irish Celtic mythology, the crow is associated with three sister goddesses, Badb, Macha, and the Morrigan (sometimes all three are identified at the Morrigan or as Badb). All three were battle goddesses who could appear at one moment as hideous hags, at the next as beautiful women, or yet again as crows. Their terrifying presence prophesied death and war. In the "Ulster Cycle" (a group of ancient Irish legends), the Morrigan as a woman appears to Cuchulainn and offers him her love. He refuses her four times. He encounters her again as the *Bean Nigh*, Washer at the Ford, where she is cleaning his own bloody armor and thus foretelling his imminent death. Cuchulainn's enmity for crows had been long-standing. He pursued otherworldly ones to destroy them and he mocked the goddess to her face. He seemed determined to defy the forecast of death. In the end, he could not escape his fate.

Cuchulainn may in fact have been a real Gaelic warrior whose history, in classic Irish fashion, turned into legend which expanded to mythic proportions. He came to be considered the mortal son of the sun god Lugh (who was himself described as a warrior). Today, Cuchulainn persists as a culture hero. His representation as a martyr willing to give his life for Ireland is used on both sides of the conflict in Northern Ireland (nationalist and unionist) to promote different ideological narratives and ideas of national unity. Nationalists see him as the greatest Celtic hero of all Ireland, while unionist paramilitaries use the image to symbolize a hero from Ulster defending his province from enemies to the south against all odds. Always on the shoulder of this hero, the crow forecasts his death.

## The Tengu Rescue Tametomo

Japan. Retold by Ellen W. Williams.

Commander MINAMOTO no Tametomo, master of the bow and Emperor Sutoku's most brazen warrior, was defeated in battle by Emperor Go-Shirakawa's men. This was in 1156 near the island *Izu Oshima*. Tametomo barely escaped with his life.

Undaunted and feeling insolent, the samurai general decided to summon his men along with his wife, Shiranui, and only son, Sutemaru, onto a ship that set out for Kyoto. Just as they began to think they were safe, a typhoon descended upon the seas and the vessel was pulled apart. Tametomo watched helplessly as Shiranui threw herself into the raging ocean in order to quell the storm. As she was flailing in the waters, an enormous sea monster rose up to devour them all, but the fearless samurai Kiheiji, holding Tametomo's infant child to his chest, clung to the creature's back and survived the attack.

Tametomo had gripped his sword and stood to commit seppuku (suicide by sword) when, suddenly, ghostly grey *Tengu*, winged crow-creatures sent by the spirit of Emperor Sutoku, appeared to save him and his men. Two Tengus, plucked the warrior out of the turbulent ocean while others plunged forward to save his men.

Thus, Tametomo was saved by the Tengu. He went on to participate in many outstanding exploits.

More than 5,700 miles separate the islands of Ireland and Japan, but battle crows can be found in the legends of each place and, while their appearances differ greatly, the birds of both countries were consummate shape-shifters. Tengu were known throughout Japan as supernatural hybrid beings. Several forms of Tengu have existed but the earliest was known as the Karasu-Tengu (Crow-Tengu), with the beak and wings of a bird and the body of a man. Author F. Hadland Davis writes: "Part crow, part human, Tengu, were believed to inhabit specific mountain areas and to create serious mischief. Minor divinities, they were supreme at the art of fencing and weaponry. They were also capable of entering a person's mind and body in order to send them on uproarious adventures."[17]

Tengu were trickster mountain demons to be assiduously avoided. They had a habit of making bad trouble—starting fires, kidnapping priests and children, luring

***The Former Emperor [Sutoko] from Sanuki Sends His Retainers to Rescue Tametomo (Sanuki No In Kenzoku O Shite Temtomo O Sukuu Zu)*. 1850–1851. Utagawa Kuniyoshi. Japan. Woodblock print, 14.25 × 30 inches. Kuniyoshi was an exceptional printmaker whose genius found its best expression in martial scenes. In this scene, the famous archer Tametomo is saved from a wrecked boat by supernatural crow-beings, *Tengu*, who lift him up just in time. The swirl of the waves competes with the wind-blown wings of the ghostly Tengu and the open jaw of a sea creature to create a moment of great drama (Photograph © 2023 Museum of Fine Arts, Boston).**

people into the woods with strange sounds, and changing form. Tengu could instantaneously appear, disappear, and reappear somewhere else.

Tengu were taken seriously because only *yamabushi,* wandering ascetics who lived deep in the mountains and who were trained in mystical arts, could overcome their power. As late as 1860, officials felt it necessary to post this official notice in advance of a visit from a *shōgun* (military commander):

> *TO THE TENGU AND OTHER DEMONS*
>
> *Whereas our Shōgun intends to visit the Nikko Mausolea next April, now therefore ye Tengu and other Demons inhabiting these mountains must remove elsewhere until the Shōgun's visit is concluded.*
>
> *(Signed) MIZUNO, Lord of Dewa*
> *Dated July 1860.*[18]

The legend of Minamoto no Tametomo, illustrated in a woodcut here, bears a resemblance to Cuchulainn's legend in that both men were historical warriors whose reputations grew and grew. Tametomo was known as a great archer and a *busho* (war lord). Stories of his prowess were well known in Japan. The artist Kuniyoshi Utagawa (1798–1861) drew from the work of Bakin Kyokutei, *Chinsetsu Yumiharizuki,* which recounted Tametomo's activities fleeing defeat in the Heiji Wars. The narrative presents Tametomo and his party of warriors as they are overcome by a storm. To calm the seas, Tametomo's wife throws herself into the waves, while an attendant saves their child by climbing onto the back of a huge sea creature. Tametomo is about to commit *seppuku* (suicide by disembowelment) when the Tengu followers of the dead Emperor Sutoku save him.

The Tengu embedded themselves one way or another into almost every type of religious practice to flourish in Japan, but their most dramatic visual representation can be seen in battle armor. Samurai warriors, who ruled the island nation from the twelfth to the nineteenth centuries, coveted expertly designed, constructed, and decorated suits of armor that would serve as functional military gear and as signs of status and wealth. The Tengu motif caught the imagination of men who wanted to replicate a mystical image, gain Tengu blessings, receive their magical powers, and scare the enemy away all at the same time.

In sixteenth-century Japan, a *Kawari kabuto* (eccentrically shaped helmet) was a fashionable accoutrement for any elite soldier who wanted to be easily identified in the chaos of the battlefield. The iron bowl protected the samurai's head while the Tengu ornamentation delivered a special kind of protection. With sharp "horns" bursting from the temples, this helmet was designed to frighten and impress. The bird-face depicting a Tengu is highlighted by the whites of its eyes. Two embossed

**Exotic helmet with Tengu mask and crows. 19th century. Japan. Iron, 8.25 × 9.5 × 11.25 inches. Whether seen on the battlefield or in a city parade, this samurai Tengu helmet would have surely intimidated any potential foe. The creation of samurai armor became a unique Japanese artform by the mid-nineteenth century (Bequest of George C. Stone, 1935. Metropolitan Museum of Art).**

crows flank its sides. They are poised in flight just behind their leader. The simplicity of the design further conveys strength.

During the Edo period (1615–1868), warriors bedecked themselves with armor that imitated the great samurai who had trained under Tengu. One of Japan's most revered samurai Minamoto no Yoshitsune (1159–1189) reportedly lived among the Tengu in the mountains and received training from the Tengu king himself, Sōjōbō. By this time, however, samurai armor was used more for ceremonial purposes to show high social rank than for combat. The artform required specialized training in a number of areas, from blacksmithing and metalworking to leatherworking and weaving, and took months to complete. Immense creativity and artistry were conveyed through design, embellishment, and dramatic flairs.

Ideas around the nature of the Tengu changed dramatically over time. At first, they were thought of as birds of prey, then as men with bird heads and wings, then as powerful men with long noses. They became more and more human, and were seen as disrupters of Buddhist sects, then as disrupters of society in general. Their role as mountain village troublemakers faded somewhat. Nevertheless, stories of the Tengu continue to be passed on from generation to generation and villagers will take precautions. Rice cakes are offered to the Tengu before clearing a forest or beginning a hunt, and prayers are directed to them when a child is lost.

## Mortal Men

Haida, British Columbia. Collected and translated by John R. Swanton.

When he first made human beings, Raven said they would look like stones, and never die; but when a small wren (Tc'a'tc'a) heard it, he said "Where shall I call, if men live forever?" [This bird calls underneath the graves]. So Raven made men mortal to give this bird a place to call.

## Nâs-Ca'kî-Yêł

Tlingit, Alaska. Told to John R. Swanton by Katishan, Chief of the Kasq!aque'di of Wrangell, Alaska, in 1904.

Nâs-ca'kî̱-yêł (Raven-at-the-head-of-Nass) tried to make human beings out of a rock and out of a leaf at the same time, but the rock was slow while the leaf was very quick. Therefore human beings came from the leaf. Then he showed a leaf to the human beings and said, "You see this leaf. You are to be like it. When it falls off the branch and rots there is nothing left of it." That is why there is death in the world. If men had come from the rock there would be no death. Years ago people used to say when they were getting old, "We are unfortunate in not having been made from a rock. Being made from a leaf, we must die."

Characteristic of most of his undertakings in Northwest Coast tales, Raven here creates death almost randomly. In one case, the pleading of the Tc'a'tc'a bird changes his mind; in the other, he is simply impatient and assigns our fate to match that of the fragile leaf instead of the immutable rock. What are we to make of this? Perhaps the lesson is for us to remember that we are not categorically different or better than a Tc'a'tc'a bird, a leaf, or a rock, and from the very beginning life has always been provisional. We are part of Raven's world, one that changes all the time. Raven himself is capable of dying—over and over; the foremost world-maker is also the premier un-doer. This singular figure of great capacity and significance has his beak in just about every situation. He stimulates the world and instigates profound change. He is a trickster of consequence, but one that we can also laugh at. Hundreds of stories describe his exploits. Raven stories, in turn, form a lodestone of Native identity all along the Northwest Coast.

The Indigenous peoples of the Pacific Northwest are divided into many nations, each having distinctive and unique traditions. At the same time, they share an ecosystem rich in natural resources and an oral tradition that moves and morphs across the land. And Raven's own peregrinations link them all. Raven stories were told at Haida potlatches, in Tlingit cedar-plank homes, during Tsimshian feasts, at the Heiltsuk (Bella Bella) Hamatsa rituals, around Nuxalk (Bella Coola) fires, and on the beach in Tillamook. His presence was felt everywhere.

John R. Swanton (1873–1958) arrived on the Northwest Coast in 1900, a small, shy, sensitive young ethnologist. He had trained at Harvard under the eminent anthropologist Franz Boas and thus threw himself into a cultural immersion on Haida Gwaii, an archipelago off the coast of Canada, ultimately creating the first comprehensive documentation of Haida myths and stories. Swanton sat with a translator for hour after hour recording and transcribing each story phonetically phrase by phrase. At the time, Haida language had not been written down and Swanton's work proved invaluable to future generations. He believed that stories should be recorded exactly as they were told in the original language with native vocabulary and syntax. He valued the artistic choices of the storyteller, asserting that no one story existed immutable, and opted to provide minimal commentary of his own, suggesting the material could be assessed on its own terms. A few years later, Swanton turned his skills to the Tlingit of southeastern Alaska and the west coast of Canada where he recorded and translated a large number of their myths and songs. His work then led him to the Indigenous cultures of the American South. Swanton's work influenced the course of American anthropology. He was at the forefront of those who turned away from evolutionary theory and was one of the first to utilize what became known as "ethnohistory" (the use of historical records to inform an understanding of living cultures). He deeply believed that there were many kinds of truth.

**Raven rattle. Ca. 1850. Attributed to Albert Edward Edenshaw, Haida. Wood, pigment, glass beads and vegetal fiber, 4.75 × 13.5 × 3.5 inches. Raven rattles were used in pairs by tribal chiefs during important ceremonies. They were shaken rhythmically to increase the drama of an oration. This rattle shows a shaman initiate with a bear or wolf head on the back of a raven, facing another raven. He is drawing inspiration, power, and wisdom from the animal world through his tongue. In the larger raven's mouth is a red disk, a reference to its role in bringing daylight to the world (The Charles and Valerie Diker Collection of Native American Art, Gift of Charles and Valerie Diker, 2019. Metropolitan Museum of Art).**

## Crow Makes a Woman [and Initiates Death]

Barkindji. Far West New South Wales, Australia.

Crow saw Walbu (bell-bird) making a woman out of mud.

"What are you making?"

"A woman."

"But she'll be no good. She'll hop like you and spill water when she's coming from the water-hole."

He went away.

Later he took a drink at the water-hole. He saw his reflection in the water.

"How beautiful" he said. He made it into a woman.

Later she married. Crow came to see her. She was alone.

"Where's your husband?"

"He's dead, but he's coming back to life again in three days."

"He can't do that." said Crow.
He went to the grave. The spirit was just coming out.
"Go back!" he said, "die altogether!"
He was after widows.

Competition between Crow and other creatures lies at the heart of many Aboriginal stories in Australia. Even though they are supernatural beings and powerful actors (creators and destroyers), these mythic beings behave very much like ordinary human beings—lustful, jealous, petty, resentful. Their interactions also mirror that of the natural world. For instance, in the stories, Eagle-man routinely has something Crow-man (*Wahn*) wants. In nature a similar situation exists, eagles mostly do the killing and crows try to edge in on the meal.

In this story we learn something about death—it was not always a final event. When Crow runs into the woman he made from his own reflection, she tells him that her husband is dead but will be coming back to life soon. From the tone of the narrative, this does not appear to be an unusual occurrence. But Crow won't stand for it; he wants death to be permanent so he can marry the widows. He commands the spirit to "die altogether!" Crow himself dies over and over in Australian mythology. Somehow, he is able to resurrect himself, much as a shaman does.

The Murinbata people of the northeastern Victoria River District relate an amusing argument between Crow and Crab about the best way to die. Crab crawls into a hole, sheds her shell, and waits for a new one to grow (thus resurrecting). Crow waits and watches her activities but gets impatient and declares that he has a better way. He rolls back his eyes and falls over dead (presumably permanently). The Murinbata people perform a ritual dance that recreates these two types of death. It portrays Crow's way as the better way.[19]

Crows are associated with death in other contexts. Among the Noongar people of western Australia, the birds are believed to help carry spirits of the dead across the western sea to the afterlife. Mudurup Rocks, just south of Perth's Cottesloe Beach, was regarded as sacred to the crow and the place where the spirits left the land for their final journey.

# 7

# Magic and Medicine

Crows and ravens have many powers. They are first and foremost existential birds—thought to foretell death but also create life. Maybe their other powers derive from being in between, neither this nor that, here nor there. After all, the crows that visit me daily are songbirds without a song. They are carnivores but not true hunters. They patrol the skies but can be found scratching around my lawn. They are my neighbors but not of much practical use to me. Because the birds behave in an elusive way, we are inclined to think them magical. Wild crows are notoriously difficult to trap and band. Zoologist Carolee Caffrey says she succeeds on average only one out of every three attempts; probably because the birds learn danger quickly.[1] Bernd Heinrich says about the ravens he studies: "What at first impressed me most about these birds was their extraordinary shyness,.... Most ravens in Maine are so shy and sharp-eyed that they immediately fly away if you so much as stop to look at them from half a mile away."[2] Their nests are often impossible to find. Another biologist remarked: "Ravens are the most exasperating birds to work with ... they are masters at concealing nests, and in general, smarter than man."[3] They have another special skill as well—the ability to hide their food very successfully.

## *Field Notes: Caching*

Over 200 bird families exist in the world (North America hosts 80). Of these, 15 bird families are known to cache (hide) their food. Corvids are pretty good at this sneaky practice. Usually starting late summer into early winter, the birds will gather surplus food items—seeds, nuts, and dead insects and animals—to hide for future use. A cache might be pieces of corn pushed into a rock crevasse, a dead mouse buried in the soil, or a pile of seeds wedged into the pocket of a tree branch.

Crows and ravens are known as scatter-hoarders. They will use hundreds, even thousands, of locations in which to cache their food treasures. They are capable of stuffing many seeds into a bill or esophagus. They can also hold a large worm under the tongue and fly to a different spot. The birds will use their mandibles to widen a dirt hole then disgorge their goods, or they will alight on a tree and carefully place each item in its proper position. They then proceed to select bits of earth, leaves, or rocks that will best cover and camouflage the hiding place. Dr. Lawrence Kilham recorded one observation: "A crow on March 3, 1984, tore up the body of a frog held down with its toes, then walked in four directions to store portions of it before flying

**A carrion crow in the U.K. is too busy caching its treat to notice the human with a camera nearby. Crows have a remarkable ability to remember where they store their food. Some will return to a hiding spot to inspect or add more to it (Photograph courtesy of Dr. Africa Gómez).**

off with the carcass."[4] They never use the same location twice, but seem to be able to remember where most everything is anyway. Kilham further commented: "Female A gave begging calls while incubating on March 14, 1984. When no crow came to feed her, she flew to an air plant at the same height as her nest and 12 meters away, took out the remains of a frog, which she ate for two minutes, and then returned to her nest."[5]

Short-term memory is retained in the brain's hippocampus. Among birds who cache, like the crow and raven, the hippocampus is proportionally larger than in other birds. Ravens are known to remember cache locations for at least two weeks. Dr. Richard Elliot, with a team of Canadian Wildlife Service scientists, monitored ravens patrolling a colony of murres on Digges Island in Hudson Bay. He watched as they took eggs, eating some and caching others. Elliot believes one of the ravens he was watching hid more than a thousand eggs and retrieved all of them.[6] Dr. Bernd Heinrich says: "Their minds seem to be like a chalkboard. They have the capacity to register many locations on it, but with each retrieval they 'erase' that site from further consideration."[7] As far as crows go, they seem to remember for even longer periods of time. Nicola Clayton at Cambridge University reports that the crows she studied weren't just remembering where they'd hidden their food; they were remembering when they'd hidden it *and* whether the food was still fresh.[8] Her study is supported by another on western scrub jays (corvids), who were documented overlooking potentially degraded food caches in favor of those that were fresher. They were remembering how long ago they had buried the food.[9]

These birds also seem to be aware of who is watching them and who is listening, and they adjust their behavior accordingly. If a crow sees a non-relative observing it, it will likely move its cache. If another bird can hear but not see what that crow is doing, it will silence its movements and use only hiding places that don't make noise.[10] It knows that an onlooker will remember where the treasure is hidden. A crow is both a hider and a stealer, and knows others have good spatial memories too.

Ravens are particularly attentive cachers. They will use trees or rocks to hide themselves from view when they cache. They will also travel great distances to place food in locations that no one else would see. This is particularly true if they are in a group setting where every bird is competing (a pair or a lone bird will cache closer to a food source presumably because no one important is around to notice them).[11] Dr. Heinrich spent time studying caching behavior. He suggests that ravens have remarkable mental awareness, stating that they "judge their competitors on the basis of what they remember them paying attention to. They then attribute to the competitors the capacity of knowing, and they integrate that knowledge together with dominance status into strategic decisions for making and retrieving caches."[12] He adds: "All of my observations of raven-caching behavior indicated more flexibility than had ever been observed in any other animal."[13] Apparently, they are even capable of telling a lie—they will make a false cache to fool those around them and then hide the food elsewhere.[14]

The discovery by another animal of a hidden morsel of food might seem a bit magical to that animal. As a practice, caching helps the environment. Caches support hungry visitors who happen upon them, and uneaten stores will sprout into new plants and trees that rejuvenate the habitat in a way perfectly suitable to the birds. It's not magical at all but it is mysterious. The elusive maneuvers maintained by the crow and raven enrobe them in the magic that seems to attach itself to mystery. Particular people, the shaman and the witch, have been privileged with access to the bird's supernatural abilities; the shaman's main job is to heal the ill while the witch invokes its powers for less benevolent purposes. But, as we will see, sometimes the bird accesses its own magic for more amusing outcomes.

## *The Witch of the West*

Crows and ravens have long been associated with witches although the origin of the connection is unknown. The bird is sometimes identified as a witch's "familiar" (a spirit animal who serves as assistant). Today's Halloween stores are lined with images of witches with crows on their shoulders. Witches are also portrayed flying on a broom surrounded by ravens. And, in the famous 1662 witch trial of Scotland's

**Baba Yaga. 1900. Ivan Bilibin. Illustration for "Vassilisa the Beautiful." Baba Yaga appears in many Slavic folktales as a terrifying old woman who flies the Earth in a mortar and wields a pestle as a weapon. Her magical powers are extensive—the simple wave of her handkerchief will summon a raging river. She is likely to try to cook you up and six crows perched in the trees are waiting for some leftovers (Artepics/Alamy Stock Photo).**

Isobel Gowdie, young Isobel confessed to the jury that she could become a corvid. Indeed, many folktales describe a witch transforming into a crow. This was the case for Baba Yaga, a well-known character in Slavic folklore.

Baba Yaga was the quintessential witch. She was hideous looking and wicked. She lived in the forest in a hut built on moving chicken legs, where she cooked and ate the children she had captured. Her fence was decorated with human skulls. She rode through the air and wiped away her traces with a broom. Although Baba Yaga was frightening, she had an alternate side. In a number of tales, she is a change agent, helping a character complete a quest or escape a situation. Baba Yaga is best known through a story called "Vassilisa the Beautiful" in which she frees a young woman from the grips of a cruel stepmother. In one version of this story, Baba Yaga eventually turns into a crow.

In L. Frank Baum's famous book, *The Wonderful Wizard of Oz*, published in 1900, the Wicked Witch of the West controlled, among other things, a flock of 40 crows. The crow and raven were further useful to our film industry's storytelling when Disney Studios created Diablo (later renamed Diaval) to be a raven servant and loyal friend to the evil witch Maleficent. Most recently, film director Joel Coen used the crow motif for his trio of witches in a dramatic 2021 black-and-white, stripped-down rendering of the Shakespearean play *The Tragedy of Macbeth*. Actor Kathryn Hunter (playing all three witches) and Coen started out by discussing the characters: "We agreed we didn't want to do witchy, witchy, pointy nosy, nasal voices kind of witches. But Joel, because he has such an extraordinary imagination, very quickly gave a starting point, which was completely different. He said, 'I think they're crows. They're like crows. And sometimes they're like standing stones. And sometimes they're women.'"[15] In her performance, Hunter was able to absorb the eeriness of a crow and the evil nature of the witch.

British painter John William Waterhouse (1849–1917) created a languid,

***The Magic Circle*. 1886. John William Waterhouse. England. Oil on canvas, 72 × 50 inches. Like many people of the Victorian era, Waterhouse was fascinated with the exotic. Here he presents a Middle Eastern enchantress at work. Her bird "familiars" wait for her instructions (The Tate Gallery [Presented by the Trustees of the Chantrey Bequest 1886], London. Photographic Rights © Tate 2018, CC-BY-NC-ND 3.0).**

moody work in 1886 that perfectly conveys the unsettling aura surrounding an enchantress and her ravens. Magic and prophecy were common themes in Waterhouse's art. Here he sets out a dark scene in a barren landscape. A lithe, beautiful woman with long hair and an oriental or gypsy dress draws a delicate circle of fire around herself with her sharp wand. Smoke rises in a thick column from the brazier. Outside the magic circle congregate the ravens. On the left, one stands on a dead animal, signaling its commerce with death. The birds appear to be watchers. They stand by, or bow, and do not flap or fly. They are waiting for something. We are left to make our own determination about what the mysterious woman is doing but all the images found in the painting could be interpreted as being symbolic of witchcraft.

## A Shaman's Magic Words

Aivilik, Arctic Circle, NW Hudson Bay, Canada. Collected and translated by Knud Rasmussen.

When one sees a raven fly past, one must follow it and keep on pursuing until one has caught it. If one shoots it with bow and arrow, one must run up to it the moment it falls to the ground, and standing over the bird as it flutters about in pain and fear, say out loud all that one intends to do, and mention everything that occupies the mind. The dying raven gives power to words and thoughts. The following magic words, which had great vitalizing power, were obtained by Angutingmarik in the manner above stated:

nunamasuk
nunArzuamasuk
ubva mak·ua—
saunErjuit siLArju p
qAqitorai—
pArqitorai—

he-he-he.

to ·$^{r}$ j$^{r}$ A·rzuk
to ·$^{r}$ j$^{r}$ A·rzuk
udludlo
avatijnut
audlArit
patqErnagit
u$^{w}$ ai-$^{uw}$ ai-$^{uw}$ ai!

*Translation:*

Earth, earth,
Great earth,
Round about on earth
There are bones, bones, bones,
which are bleached by the great Sila
By the weather, the sun, the air,
So that all the flesh disappears,
He – he – he.

Spirit, spirit, spirit,
And the day, the day,
Go to my limbs
without drying them up,
Without turning them to bones
Uvai, uvai, uvai.

Angutingmarik was a respected shaman of the Aivilik-speaking people residing on the shores of Canada's Hudson Bay. Ethnologist Knud Rasmussen worked closely with this unique man as well as many other *angákut* (shamans) during his travels on the Fifth Thule Expedition (1921–1924) across the Arctic from Greenland west to Siberia. Twenty-seven reports were generated from this expedition. Two of the stories Rasmussen collected have already been noted in Chapter 1 and 2. Here we enter the world of the spiritual mediator.

Rasmussen wrote that shamans carried out many roles. They not only had to forecast the weather but also ensure good weather. They had to be able to fetch game (*nak'aivaglune nErgutinik manisaidlune*, meaning: "they must be able to fall down [to the bottom of the sea] in order to bring to light the animals hunted.").[16] Angákut also visited the Lands of the Dead (one was under the sea while another was in the sky) to retrieve lost or stolen souls. But most importantly, they were called upon to cure the sick.

It was said that in the beginning there were no shamans on Earth until there came to be a period of sickness and starvation. The first shaman was a man who dove down through the earth to confront the Mother of the Sea Beasts and bring game back for humankind. "Afterwards, the shamans extended their knowledge of hidden things, and helped mankind in various ways. They also developed their sacred language, which was only used for communicating with the spirits and not in everyday speech."[17] Becoming a shaman (both men and women could be take up the job) took time and special training from a mentor. Helping spirits were enlisted first: the mentor had to "remove the pupil's soul from his eyes, brain, and entrails and hand it over to the helping spirits so that they will know and recognize him."[18] Then the pupil's own particular "light" had to be secured.

> The first time a young shaman experiences this light, while sitting up on the bench invoking his helping spirits, it is as if the house in which he is suddenly rises; he sees far ahead of him, through mountains, exactly as if the earth were one great plain, and his eyes could reach to the end of the earth. Nothing is hidden from him any longer: not only can he see things far, far away, but he can also discover souls, stolen souls, which are either kept concealed in far, strange lands or have been taken up or down to the Land of the Dead.[19]

The new angákut then underwent training and purification. He met with answering spirits and endured time alone in the wilderness.

> The last thing a shaman learns of all the knowledge he is obliged to acquire, is the recitation of magic prayers or the murmuring of magic songs, which can heal the sick, bring good weather or good hunting. One can practise magic words simply by walking up and down the floor of one's house and talking to oneself. But the best magic words are those which come to one in an inexplicable manner when one is alone out among the mountains. These are always the most powerful in their effects. The power of solitude is great and beyond understanding.[20]

We have read how Angutingmarik chased down a raven to obtain magic words from its dying mouth. With all his powers assembled, he might have offered the skin of a raven to a newborn child. A baby's first garment was sometimes made as a cap or as a dress of raven skin, with the feathers on the outside, to give the child the power to discover game easily.

## Word Charm

Malaysia. Collected and translated by Walter W. Skeat.

With a blind crow as his guide,
The giant demon, Si Adunada,
Carries his weapons slung
Over his shoulder
With back bent double.
Salampuri is the name of his sekin
Silambuara the name of his k'ris
The insignia of the Demon Rama.

Malaysian jungles were once terrifically feared places. Although they were rich in resources, they were so filled with supernatural powers that one seldom ventured into the interior alone. A Specter Huntsman, for instance, roamed the forests with several ghostly dogs. He was a being to be avoided; simply crossing paths with him brought fatal choleraic symptoms. Numerous charms to neutralize his influence were employed by the medicine men (*pawangs*) of the area. Like the shaman of the Arctic, the pawang of a Malay village held great status. He developed a working relationship with a familiar spirit (in this case a hereditary family spirit) through which he was able to deal with wild and dangerous influences. He acted as spirit-medium, gave oracles, could affect the weather, and commanded political influence within village life.

Adunada, mentioned above, was the Specter Huntsman's son. He too was lethal. The charm we read about subdued him, his supernatural characteristics ("guided by a blind crow"), and his royal weapons the sekin (a straight-edge knife) and the k'ris (a wavy dagger). Repeating this mantra over and over was a way to show him respect and to indicate that one was his follower, and therefore to be spared his wrath. Charms such as this might have been written down and worn in clothes, or chanted throughout the day, or used at those healing ceremonies when an illness could be traced to an encounter with the Specter Huntsman. Every tactic was needed to keep the dreaded Huntsman and his son at bay.

The English anthropologist Walter William Skeat (1866–1953) began his ethnographic work when the field of anthropology was just finding shape. As much as any of its early students, Skeat was buffeted by the field's turbulent philosophical winds. He preferred to record what he heard and saw rather than provide commentary, but his perspective as an administrator in the British Colonial Service no doubt affected his impressions. Skeat collected most of his data from villagers living in the Kuala Langat district on the west coast of Peninsular Malaysia. The area was at that time home to about 3,000 Malay who lived in remote settlements of 100–200 people.

## MIDĒ PICTURE-SONG #8

Ojibwa (Anishinaabe), Canada.
Collected and translated by Howard Norman.

Sure!
all the clouds are dark
crow
wants us (all) to dress like him
must be time for it

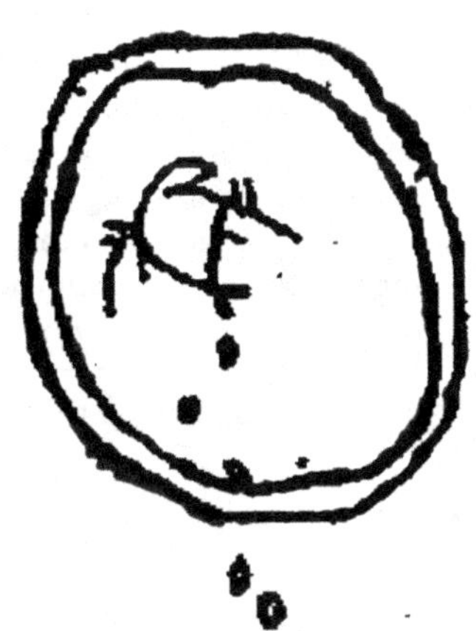

**Crow mide. (drawing by author).**

The Ojibwa tribal peoples of southern Canada and north central USA (from Minnesota to North Dakota) call themselves Anishinaabe, "first" or "original" people. They have always been profoundly connected to the land that surrounds them and their mide picture-songs are created out of the interwoven nature of their lived experience. No firm distinctions are made by the Anishinaabe between animals and humans as each shares the other's essence. Objects, events, or activities are also porous, representing different things on different occasions. Theirs is a fully integrated, fluid world where the transformation of any form of being is possible and all forms of being can be channels for spiritual agency. You never really know what you are looking at when you come across a bear—it could be a person as a bear, it could be a spirit as a bear, or it could a regular bear. For this reason, the Ojibwa people turn to animals as pivotal symbols and keepers of Ojibwa values.

An Ojibwa mide song is a sacred thing. Acquired through powerful dreams, composed by an individual healer (a *mide*), and performed during ceremonies, mide songs lie at the heart of the Grand Medicine Society or *Midewiwin* religion. They are songs of healing and empowerment, of history and memory, of moral and spiritual lessons. Picture-songs were written down by a mide on the soft inner surfaces of birch bark using a sharp nail or piece of bone. They took form as a series of pictographs to lead the healer through a song and ceremony. These objects were used to make things happen but were considered so powerful by themselves that they were often placed under the care of a guardian.

Paul Buffalo, a mide from Minnesota, spoke in the late 1960s and early 1970s about the work of medicine doctors.

> Grand Medicine is more a doctoring work. It gives life to the Indians when they respect one another, and respect the birds, respect the animals, respect the nature of the timber outdoors. It gives them life because these timbers and things are given to them for medicine … they [the mide] empower the medicine. After it's empowered, the medicine is used by this man, or by anybody.… They grind up herbs. Yea. They grind them up and they put them in a small buckskin bag—a "medicine" or "luck" bag. Some people call it a "luck bag," in English.… It's all blessed by the powerman. They have a drum and they drum and sing. They sing for empowerment and everything … they pound the drums and sing, just like the songs I sang for you. They pound a drum and they dance. They have medicine and they dance around carrying their medicine in their fur hides: mink hides, skunk hides, beaver hides.… They sing a spiritual song. There's an answer to everything. When this guy sings the song, he sees he will get an answer. It's a spiritual song. It's a medicine song.… That sound at the end of the song is when they get answers. It's a medicine song. They're talking to the spirits. It's a spiritual song. Those songs are nothing to fool with either. The Medicine Men get an answer by them. Birds carry the message. Birds carry these songs. When you see the bird around, he's answering. Something's going to happen.[21]

In Picture-Song # 8, a crow appears inside the sky, or a cloud, from which he is releasing precious drops of medicine. We cannot know the true meaning of the song's words—they were specific to the singer, the context of the performance, and were the provenance of his secret magic—but we can feel the power of the crow. We are ready to dress like him. We are ready to receive his medicine.

A whole birchbark pictograph should never be shown. It is too sacred. Paul Buffalo warns:

> The Midewiwin had birch bark writing, but we just don't say much about that—to anybody!!.… To interpret those birch bark writings you have to be in their session when they make them. A lot of stuff backfires on a lot of people when they don't stop, look, listen! A lot of them aren't here because they didn't respect that fence at the *Midewiwin* gatherings. And they didn't respect where it's fenced on those writings. The Indians said that fence is a stop sign. Don't go against it, in any way! That's what I believe in.[22]

## Raven Magic

Morocco. Recorded by Edward Westermarck.

I have heard that the raven was once a blacksmith; and this may be the reason why it is so difficult to shoot a raven—it sees the bullet and flies away (Aglu [tribe near the Atlantic coast]).

To shoot a raven may also be attended with evil consequences. I was told of two men among the Bni Msa'uwar [tribe near Tangier] who did so and whose guns burst, and similar things are said to happen among the Ait Wäryâg'er [northeast Morocco]. In Andjra [east of Tangier] I heard of a man who became mad because he killed a raven.

[But] if a woman is losing her hair, a raven is killed and charred and her head is smeared with the powder mixed with honey (Andjra) or with water or oil (Ait Waráin [interior Morocco]); this will give her a good growth of black hair, like the feathers of the bird.

A childless woman who is desirous of offspring tries to get hold of a raven to kill and drink its blood, warm as it flows from the body (Aglu).

The raven is eaten as a cure for syphilis (Ulâd Bu'āzîz, Ait Ndēr, Iglíwa)....

The Ait Sáddēn [central Morocco] drink the gall [of a raven], still warm, as a safeguard against bullets, and also to remove *atqqaf* caused by witchcraft.

The neb of a raven is hung round the neck of a little child as a protection against the evil eye, and for the same purpose a raven's foot is tied to the churn (Ait Waráin).

Charms are written with the blood of a raven (Hiáina [central Morocco]).

At the turn of the twentieth century, Edward Westermarck (1862–1939) set out to visit the dozens of Berber tribal groups living across the expanse of Morocco. In the preface to his 1926 *Ritual and Belief in Morocco*, he writes: "I contemplated going to the East to study both civilised and savage races. I sailed for Morocco in 1889—and never went farther."[23] He made twenty-one trips over the course of twenty-eight years and, although he made a career writing about such topics as the history of human marriage from a Darwinian perspective and the development of moral ideas around the world, he was captivated by the convictions of Morocco's Berbers.

From Westermarck's accounts, we learn that the raven was thought to have had concrete magical properties. It's physical body and blood were used for medicinal purposes and as talismans for good luck. The attributes extended to the bird are not metaphorical or symbolic or "spiritual" in any way, but specific and practical. And the beliefs appear to have been widespread across many different Berber groups.

Berbers are the original inhabitants of Morocco. The term derives from the

Roman word for "barbarian," but they call themselves *Amazigh*, meaning "free people." They inhabit a remote, dry, rocky area created by the Atlas Mountain chain which extends from the Atlantic coast hundreds of miles inland toward the Algerian border. When Arabs invaded North Africa in the seventh century CE, they "Arabized" and transformed the existing culture with the result that all Berbers are now Sunni Muslims. Nevertheless, they attach great importance to their tribal culture, language, and practices.

When Westermarck wrote his book on the beliefs of the Berbers, he was careful to emphasize his methodology, saying: "I have made it a stringent rule not to accept statements of others than natives of the country.... I have further made it a rule not to use information given me about a tribe by members of other tribes. I have also been in the habit of repeating to my informants their statements in full so as to avoid all misunderstanding; and I have occasionally tested their trustworthiness by deliberately misrepresenting their statements, but in such cases they have never failed to correct me."[24] Westermarck credited his assistant by name underneath his own on the book's title page. This was rarely done at the time. And he was one to set a precedent, establishing himself as a pioneer in immersive fieldwork during the early years of anthropological endeavors. His detailed records have been consulted by later scholars over and over again.

## The Raven-Stone

Prussia. Described by Walter Kelly.

A man may make himself invisible whenever he pleases if he is possessed of a "raven-stone," a talisman which is procured in New Pomerania in the following manner. When you have discovered a raven's nest you must climb the tree, and take your chance that the parent birds are at least a hundred years old, for otherwise you will have your trouble for nothing. You are then to kill one of the nestlings, which must be a male bird, and not more than six weeks old. Then you may descend the tree, but be very careful to mark well the spot where it stands, for by-and-by it will become invisible, as soon as the raven comes back, and lays a raven-stone in the throat of its dead nestling. When it has done this, you may go up again and secure the stone.

Northern Europe was at one time swarming with superstitions; some of them swirled around the crow and raven. In 1863, Walter Kelly published *Curiosities of Indo-European Tradition and Folklore* that included this bizarre method of making oneself invisible. A similar method is described in L. Lloyd's 1854 *Scandinavian Adventures* whereby an invisibility-producing raven-stone, *Korp-sten*, could be obtained from inside the body of a raven and then swallowed.[25] The stone was also thought to heal tumors, swellings, sores, and a great many other things.

In 1502, Camillus Leonardus wrote *Speculum lapidum* (The Mirror of Stones), in which he gave instructions about what you should do with *Corvia*, or *Corvina*, a stone of reddish color.

> On the first of April, boil the eggs taken out of a crow's nest, until they are hard, and being cold let them be placed in the nest as they were before. When the crow knows this she flies a long way to find the stone, and returns with it to her nest, and the eggs being touched with it, they become fresh and prolific. The stone must be immediately snatched out of the nest. Its virtue is to increase riches, to bestow honours, and foretell future events.[26]

We can only hope that no one really tried to secure a raven-stone.

## Come, Raven!

Acoma Pueblo, New Mexico. Collected and translated by Matthew W. Stirling.

Come, Raven!
You represent the whirlwind,
sweep away from us
this disease and all diseases and sadness
You are the one
who has the real power to do this.

When misfortune and illness get out of our control, we beseech the higher powers to help us. Indigenous peoples of North America have long turned to the crow or raven for aid. At the beginning of their time on Earth, the Acoma people of the New Mexico area summoned Raven before embarking on the first part of a great migration. They were in need of his help and their call to him here is a small portion of an extraordinary origin myth.

In the myth, the first humans (two sisters) emerge from an underground place to populate the world with life—plants, trees, birds, animals, mountains, and great spirits. The sisters multiply the human race, and they develop towns and rituals. After a series of severe illnesses, medicine men decide to consult with the *katchina* (spirit) powers to learn all there is to know about curing and how to imbue masks and costumes with the life force of the katchinas. But sickness comes again, with much death, and the medicine men cannot find a cure. The people conclude: "I guess our mother Iatiku does not want us to live here any more," and they begin a migration. Before setting off, the *chaianyi* (medicine men) make a sand painting on the ground, representing four mountains.

> When the people came [to the sand painting], they were to cross these four mountains and valleys and thus put the sickness that much more in their rear. All the people that came walked over, stepping on a mountain and valley in turn. The two chaianyi on the south had

> their two feathers, and would brush off the sickness of each person as he approached. The chaianyi would say, as they brushed them off, "Come, Raven! You represent the whirlwind, sweep away from us this disease and all diseases and sadness. You are the one who has the real power to do this." They would repeat this for each person that approached to cross the sand painting.[27]

The ritual continues. Their appeal to Raven works and the people travel on for a long time. They settle. And the story continues.

In 1928, Matthew Stirling (1896–1975) recorded the magnificent Acoma myth from a Pueblo man named Day Break (1861–1948), or Edward P. Hunt as he was later called, while Hunt was visiting Stirling in his office at the Smithsonian Institution in Washington, D.C. The interviews, including sacred songs, were recorded on wax cylinders. Much controversy surrounds the conveyance of this material. Day Break claimed to have been initiated into several secret societies whose ritual knowledge is closely guarded and rarely revealed. He broke with his people's codes of conduct and was exiled from the community. As for Stirling, he did nothing illegal but he never requested permission and never acknowledged his sources in his publication *The Origin Myth of Acoma and Other Records*, simply describing Hunt and his sons as "a group of Pueblo Indians." With these omissions, Stirling effectively erased the people of the story. The origin myth of the Pueblo of Acoma belongs to the Pueblo people. It is an exuberant account of their beginnings, which was epic in nature, and stands as one of the most detailed origin narratives to educate outsiders.

## The Battle of the Birds

Scotland. Told by John Mackenzie, Fisherman.

There was once a time when every creature and bird was gathering to battle. The son of the king of Tethertown said that he would go to see the battle and that he would bring sure word home to his father the king who would be king of the creatures this year. The battle was over before he arrived all but one (fight), between a great black raven and a snake, and it seemed as if the snake would get the victory over the raven. When the king's son saw this, he helped the raven, and with one blow took the head off the snake. When the raven had taken breath, and saw that the snake was dead, he said, "For thy kindness to me this day, I will give thee a sight. Come up now on the root of my two wings." The king's son mounted upon the raven, and, before he stopped, he took him over seven Bens, and seven Glens, and seven Mountain Moors.

"Now," said the raven, "seest thou that house yonder? Go now to it. It is a sister of mine that makes her dwelling in it; and I will go bail that thou art welcome. And if she asks thee, 'Wert thou at the battle of the birds?' say thou that thou wert. And if she asks, 'Didst thou see my likeness?' say that thou sawest it. But be sure that thou meetest me

**The king's son saves the raven. 1910. Henry J. Ford. Illustration for "The Battle of the Birds," in Andrew Lang's *The Lilac Fairy Book*. "The Battle of the Birds" is a popular Scottish tale that has been told hundreds of times with many variations to the story. Something about the theme of magical good luck following good deeds is irresistible to both story tellers and listeners (Chronicle/Alamy Stock Photo).**

to-morrow morning here, in this place." The king's son got good and right good treatment this night. Meat of each meat, drink of each drink, warm water to his feet, and a soft bed for his limbs.

On the next day the raven gave him the same sight over seven Bens, and seven Glens, and seven Mountain moors. They saw a bothy [cottage] far off, but, though far off, they were soon there. He got good treatment this night, as before—plenty of meat and drink, and warm water to his feet, and a soft bed to his limbs—and on the next day it was the same thing.

On the third morning, instead of seeing the raven as at the other times, who should meet him but the handsomest lad he ever saw, with a bundle in his hand. The king's son asked this lad if he had seen a big black raven. Said the lad to him, "Thou wilt never see the raven again, for I am that raven. I was put under spells; it was meeting thee that loosed me, and for that thou art getting this bundle. Now," said the lad, "thou wilt turn back on the self-same steps, and thou wilt lie a night in each house, as thou wert before; but thy lot is not to lose the bundle which I gave thee, till thou art in the place where thou wouldst most wish to dwell."

The king's son turned his back to the lad, and his face to his father's house; and he got lodging from the raven's sisters, just as he got it when going forward. When he was nearing his father's house he was going through a close wood. It seemed to him that the bundle was growing heavy, and he thought he would look what was in it.

When he loosed the bundle, it was not without astonishing himself. In a twinkling he sees the very grandest place he ever saw. A great castle, and an orchard about the castle, in which was every kind of fruit and herb. He stood full of wonder and regret for having loosed the bundle—it was not in his power to put it back again—and he would have wished this pretty place to be in the pretty little green hollow that was opposite his father's house; but, at one glance, he sees a great giant coming towards him.

"Bad's the place where thou hast built thy house, king's son," says the giant.

The story goes on and on and on. It is a true yarn in the style that Scottish Highland people were known for telling. This version of "The Battle of the Birds" was recited in Gaelic by John Mackenzie in April 1859. He was a sixty-something-year-old fisherman and dike-builder living on the estate of the Duke of Argyll near Inverary, Scotland. Known as an entertainer, Mackenzie was visited by many on a winter's night to listen to the stories gotten from his father and his father's friends. Hector Urquhart, the gamekeeper of the estate, wrote the story down in Gaelic and John Francis Campbell translated it into English. Campbell was on a mission to collect the disappearing oral culture of his people (most of whom could not read or write). He says in the introduction to his book *Popular Tales of the West Highlands*:

> The following collection is intended to be a contribution to this new science of "Storyology." It is a museum of curious rubbish about to perish, given as it was gathered in the rough, for it seemed to me as barbarous to "polish" a genuine popular tale, as it would be to adorn the bones of a Megatherium with tinsel, or gild a rare old copper coin. On this, however, opinions vary, but I hold my own, that stories orally collected can only be valuable if given unaltered; besides, where is the model story to be found?[28]

John Campbell (1821–1885) was educated at Eton and the University of Edinburgh but his heart belonged to the people he grew up with on the wild and windswept mountains and island shores of northern Scotland. When he began his life's work, the idea of collecting orally told stories was new. The term "folk-lore" had been coined only in 1846 by Englishman William Thoms. Yet Campbell was able to put together four volumes totaling 2,000 pages of stories told by the old folk. He writes:

> I cannot convey to anybody who has not experienced the extraordinary mass of stuff which is stored up in these old Highland minds. Men who cannot read a single letter or understand a word of any language but Gaelic, ragged old paupers men might pass off as drivelling idiots begin and sing long ballads which I know to be more than 300 years old, and when they have done they begin another and so for hours they go on as if they were clocks.... One story goes rambling on for a couple of hours and then it is not half done.[29]

In "The Battle of the Birds," we delight to hear how the raven shares his magical abilities with the king's son who saved him from the grip of the snake. The bird is able to carry the young man on the back of his wings. Auspicious numbers appear—three nights' lodging, seven bens and seven glens and seven mountain moors, and then a reverse trip back to each. The bequeathed castle with its orchard sounds enchanting. But the boy has opened the magic bundle too early ... and he faces his next adventure—against an angry giant.

## A Crow's Magic

Ainu, Japan. Told by Carl Etter.

When I visited Saghalien, I heard many stories about the crow gods stealing women for their wives.

In Tarantonari, Saghalien, I was talking to an Ainu named Ishikawa Iki. Said he: "In a certain village there lived an Ainu named Podanungi and his wife. The crow god lived near by. This crow god was madly in love with Podanungi's wife and wanted her for his own wife."

"Now, wait a minute Mr. Iki," I retorted. "Down in Hokkaido and up in the Kuriles where I have traveled, the Ainu have told me many stories about the gods getting off with their women. Do you mean to tell me you have had the same trouble up here in Saghalien?"

"Well," said he, "we have this story about the crow god, who, being in love with Podanungi's wife, came to Podanungi and used a strange magic upon him, putting into his mind a plan to leave home and go on a long journey in the middle of a cold winter. Even though his wife begged him not to go, he rigged up his dog sleds and set forth.

On the second day of Podanungi's absence, the crow god changed himself to look just like Podanungi and came to the latter's home, telling his wife that he had found it too cold to continue his journey, and had returned home.

She was deceived for five days until her real husband actually did decide to return. In the meantime the crow god lived with Podanungi's wife.

The story is too long to give in full, but when Podanungi had finished with the crow god, that deity was hanging to a tree with a rope tied to his neck."

Carl Etter must have laughed out loud when he heard the tale of Podanungi and the crow god come to its conclusion. Clearly, crow magic is not always effective. The story highlights a common theme regarding crow and raven—they are fallible, often incompetent, sometimes ridiculous—almost human.

Little would be known about Carl Etter if not for his 1949 book *Ainu Folklore: traditions and culture of the vanishing aborigines of Japan* and his papers now at the Smithsonian Institution. Etter studied religion in college and hoped to serve as a missionary for the Church of Christ in Japan, but the church would not support his efforts. Instead, he became a teacher at Hokkaido Imperial University and traveled the country collecting stories and taking photographs of the Ainu, their villages, and rituals. As we see in Chapters 3, 5, and 11, the crow plays many roles in the busy, porous Ainu universe.

## Crow Falls in Love With His Mother-in-Law

Told by Mrs. Angela Sidney, Tagish, Athapaskan Canada.
Collected and translated by Julie Cruikshank.

Then there's that time Crow falls in love with his mother-in-law.

He's walking along, comes to a widow and her daughter. He comes up, lay down, legs crossed, hand under his head. He talks to them nice. He's stuck on that girl. She pretty soon moves over to him.

One year later, he gets stuck on his mother-in-law. She gets sick. He wished her sick, that's why. He say, "I'll be doctor." So he tells his wife, "I know what medicine will help your mother." He tells his wife, "Send your mother to the bushes, to the meadow. Tell her to sit on the first thing she sees coming out of the moss."

Then he covers himself up with moss, all but his thing. She came there. She saw it. She sat down. She look down and saw some feathers. She figured it out. He didn't get nothing!

She goes to her daughter. "Where's your husband?"

"He went to get water."

She tells her daughter what happened. Funny thing though, she got better right away. Crow medicine I guess that is.

That's end of Crow stories. At end he gets tired of walking around. So made himself into Raven. Now he don't bother people anymore.

Another humorous story about Crow magic. It was good medicine ultimately, even though it didn't work. The trickster in Crow shines brightly in this telling, as it does in many tales from the Northwest Coast and Arctic. A trickster makes trouble, usually gets entangled in that trouble, and then sets things right again. The lessons teach us to "roll with the punches," accept and work with the forces in the universe that are out of our control, and see the comical aspects of life's events. Mother and daughter fared fairly well against Crow's silly contrivance.

Professor Julie Cruikshank recorded this story, as well as many others, from Mrs. Angela Sidney (1902–1991) who, at the time of the recording in the late 1970s, was in her early seventies. Mrs. Sidney emphasized that the tales she told were her versions of widely known stories within the Crow Cycle, which centered on Crow's peripatetic engagement with the world. They were personally important to Mrs. Sidney, who was eager to have them passed down to future generations. She wanted her name acknowledged with every version. Cruikshank notes: "The Raven stories told on the Northwest Pacific Coast become the Story of the Crow in the Yukon. Crow is the maker of the world. He goes through life having adventures, outwitting people and occasionally being outwitted."[30]

Mrs. Sidney described herself as a Tagish and Tlingit woman of the Deisheetaan (Crow) clan. She lived in a tiny Tagish community in the heart of the Yukon Southern Lakes district, formed at the Yukon River's southern headlands. The Tagish First Nation people hosted gold seekers during the Klondike Gold Rush of 1898 but the area was also positioned on an ancient fur trade route that the Tlingit people used to venture inland. The subarctic environment was difficult to survive, but the stories easily endured into the twentieth century. Thankfully, they were recorded; we can still enjoy and learn from them.

## *A Bird Artist: Jamie Wyeth*

The raven is an animal of interest to painter Jamie Wyeth (b. 1946). Wyeth has a home on Maine's craggy coast around which ravens linger, almost beckoning him to paint them. He responds to their invitation and engages in a kind of conversation. Wyeth approaches all his subjects with direct realism. He makes it his practice to notice everything and to capture the nuances and details of the figure in front of him. This is not just an image of a raven but a dynamic portrait of the bird. Devoid of background and closely framed on a huge canvas (over five feet wide), we can make out a glint in the eye, the puff of the breast, the point of the talons, and the lines in each glossy feather. This is a bird with presence, personality ... and mystery. We are right there beside it; and may be at risk of becoming one of its kind. What will we both do next?

Wyeth portrays the raven's wild beauty and at the same time hints at some deeper significance. The brilliant yellow lighting effect, like the call of the incandescent sun, only adds to the mystery. Wyeth remarked about painting: "I spend as much time with an animal or an object as I do with a person when I'm doing their portrait."[31] And about this piece: "I was alone for two months when I was doing [Raven], and I got this whole thing of, 'Is it alive with me, in the dark?' Totally freaked me out."[32] Andy Warhol viewed the canvas in 1980 at a one-man exhibit in Philadelphia and was impressed, saying: "I told him he should go even bigger."[33]

*The Raven* was the first of many "portraits" of ravens. Wyeth's intense study of their postures, habits, and wanderings was encouraged by raven researcher Bernd Heinrich who advised the artist to cultivate relationships with his avian neighbors. Wyeth has said that for him gulls and ravens represent nature's primeval wanderers who nevertheless "own" the spaces they inhabit.[34] He's attracted to their toughness. For an artist who has made a career transforming everyday landscapes and familiar animals into potent vessels for meaning, the raven is a thought-provoking subject.

And it is likely that his illustrious artist forebears would agree. Jamie Wyeth seems to channel the haunting realism of his famous father, Andrew Wyeth, and merge it with the crisp line and bold palette evident in the illustrations of his grandfather, N.C. Wyeth. He creates a portrait that is archetypal, primal, and unsettling. The bird makes us anxious. We might truly believe it could be a potent magician who, if it felt like it, could either help or harm us.

***The Raven.* 1980. Jamie Wyeth (b. 1946). Oil on canvas. 60 × 72 inches. With a slightly opened beak turned towards us and sharp claws poised to strike, the raven fixes his eye on something about to happen. The bird is guarded but ready. It is awake to the moment and fully aware (Brandywine Museum of Art, Purchased with Museum funds, 1992. © 2023 Jamie Wyeth / Artists Rights Society, New York).**

# 8

# Among the Animals

## Raven and Bear

A Tahltan tale collected by James A. Teit, 1919.

Now Raven came to the house of Grizzly-Bear, who was a strong, fierce man, and fought and ate people. When people saw him, they always ran away. Raven said, "Halloo, brother-in-law! what are you doing?" and Grizzly answered, "I am fishing." Raven said that he would help him; so he staid with him, and helped him catch salmon, dig roots, and so on. Presently he stopped the salmon from coming up the creek, and Grizzly became very hungry. One day Raven heated stones in the fire until they were red-hot, and then pretended to eat them. He took hold of them with two sticks, and passed them down in front of his body so that Grizzly could not see. The Bear thought it very funny that he should eat hot rocks. Raven said, "I am hungry, and these rocks are very sweet." At last the Bear thought he would try them. Raven heated a large stone red-hot, and told Bear to open his mouth. He told him, "You must swallow the stone at once, for, if you hold it in your mouth, it is not sweet. When it goes down your throat, you will taste it very sweet." He then threw the rock down the grizzly's throat and ran away. Grizzly became very angry, and fought and attacked everything he saw. At last he died, the rock having burned his stomach. This is why there are stripes on the inside of the stomachs of grizzly bears. Having overcome and killed the Bear, the latter had now lost his power; so he transformed him to the bear we know as the grizzly at the present day. He said, "Henceforth grizzly bears shall not be so powerful, nor so fierce, nor will they fight and kill people so much."[1]

## Raven and Bear

An occurrence documented by Jessica Teel in Alaska, 2010.

I witnessed the most amazing scene one day between two ravens and a pair of fishing bears who were successfully fishing the river for salmon. After gorging, their faces smeared with bright red blood, the two bears settled down for a long nap. The ravens had enjoyed feasting on the salmon scraps left by the bears. However, while the bears napped the ravens were out of food. We witnessed the two ravens standing face to face, talking up a storm. It was the most visually distinct conversation I've watched between two animals. The ravens were standing next to one of the sleeping bears. They bobbed their heads, talked back and forth never interrupting the other. After the conversation, one raven took flight and landed next to the other napping bear. Then both ravens walked up to the bear's faces and began jumping up and down, making noise, trying to wake up the bears. The two ravens were clearly, and in coordination, trying to rouse the bears into another round of fishing. The bears, however, with full bellies and blood fresh on their snouts, kept on sleeping through the racket.[2]

**Brown bear (*Ursus arctos*) visited by a flock of ravens. Ravens and crows have working liaisons with many animals. The bear is one such animal (Paolo Gislimberti/Alamy Stock Photo).**

## Field Notes: Kleptoparasitic Foraging

The relationship between corvids and nature's other creatures is complex and intertwined. The birds exist in a finely tuned ecosystem of competition and cooperation. Animals need each other to survive. We noted in Chapter 5, when we looked at how wolves and ravens interact, that corvids make alliances with animals who can help them. Above, we have two stories—one mythic and the other a living observation—about bears and ravens. In these two accounts, the raven shows a unique resourcefulness in trying to get what it wants. In the Tahltan myth, Raven is quite selfish; in Jessica Teel's account they are simply annoying.

Resourcefulness is also evident in the "kleptoparasitic foraging" strategies the birds engage in—meaning they are very clever at stealing. Typically, ravens follow apex predators on hunting expeditions. Once a victim is down, a raven "scout" will call out to recruit others to the carcass. They must wait for the predator to open the animal before they can begin pilfering the pieces. But the birds are also known to take the initiative. Dr. Bernd Heinrich described an instance where ravens flushed flickers, which were then killed by falcons, leaving ravens with bits to grab.[3] He believes there have been cases when ravens have worked together to lure mountain lions and grizzly bears to humans as prey.

Crows, too, will cooperate in order to secure a meal. Several will mob a larger animal to distract it while companions swoop in for the steal. "I've seen a jackdaw [a corvid] pull a rook's tail while another jackdaw comes around and [steals] its food,"[4] researcher Nicola Clayton says, adding that it's much like the classic pickpocket routine. Often one member will be the beneficiary of the theft, but at other times sharing takes place. Lawrence Kilham witnessed cooperative "sentinel" behavior among American crows. When a group of crows arrived at a feeding station he had set up, he noticed that one routinely perched on a high tree while the others landed on the ground. When the group left, the sentinel went with them. Presumably, its job was to shoo away outside intruders; hopefully, it was rewarded with a later opportunity to get at the food.[5]

Many types of intelligence are required for corvid thieving behavior. The birds need to think ahead to take advantage of a given situation. Dr. Heinrich points out that these birds have specialized to live off the food killed by animals who might just as easily kill them. They come into very close physical proximity with predators and must be able to predict their behavior. A crow needs to know what to do when a huge pair of jaws whips in its direction. A raven needs to know approximately how far the sharp claws of a bear's arm will extend. They need to be able to gauge the general mental state of the predator in their midst and determine how that animal is likely to act. In the case of a joint effort, the birds also need to coordinate their movements with one another.

And so they spend quite a bit of time testing the reactions of larger animals, teasing and provoking them. Many people have seen a raven tormenting another animal. The bird will pull an otter's tail, pluck a turkey's feather, bait a hawk into attacking. Writer George Laycock recounts that "Peter Smith, professor of biology at Acadia University, has watched lines of ravens queue up to take turns yanking the tail feathers of bald eagles on the ground. After giving the tail a vigorous tug, the raven would leap back from the danger zone and go to the end of the line to await its next turn. 'How long this game goes on,' Smith told me, 'depends on the reactions of the eagle.'"[6] The risky behavior ultimately supports survival, for the birds learn just how far they can go and what distances are required for their safety.

The talent for stealing even extends to the human world. Ravens have been photographed pulling up fishing lines at ice holes to make off with juicy fish. Pet crows are known to steal rings, watches, and silverware. Both crows and ravens have several methods for opening garbage can lids, and biologists at Yellowstone National Park claim that ravens have perfected a talent for undoing the Velcro fasteners on snowmobile storage compartments to dig into the goodies inside.[7] We're just another species whose boundaries need testing.

Hundreds of stories exist around the world that relate how a crow or raven interacts with another species. In most cases, the birds engage in trickery; sometimes they win and sometimes they lose out. Two fables are highlighted in this chapter to represent the profound resonance of a particular kind of story—the parable—where animal behavior stands in for human nature. We look at how the crow and raven, in concert with other animals, are used to show particular human predilections and what lessons are meant to be drawn from their behavior. These two fables have been remarkably enduring. They each traveled across lands, cultures, and time largely intact.

## Brotherhood

Arabic. Abridged fable from *Kalila wa Dimna*.

The king said to the wise man: I have heard the fable of the two friends whom the false trickster separated. Now give me a fable concerning sincere friends—how the beginning of their friendship came about, and how they profited, each of them from the other.

The wise man said: The intelligent man thinks nothing equals sincere friends; for friends are of the greatest help in securing benefits and of the greatest consolation in misfortune. As an example, there is the fable of the crow, the ringdove, the mouse, the tortoise, and the gazelle.

The king said: How was that?

The philosopher said: They say that there was in a certain land a place full of game in which hunters used to hunt; and in this place there was a large tree with great branches covered with leaves. In it was the nest of a crow.

One day while the crow was on the tree, he saw a hunter approaching the tree, ugly in appearance and of evil state. On his shoulder he carried a net and in his hand a staff. The crow was frightened by him and said: Assuredly something, either my destruction or the destruction of someone else, has brought this man to this place, and I shall remain until I see what he is going to do.

The hunter approached, spread his net and scattered upon it his grain, and hid himself in a place nearby. He waited only a short time until a dove which was called 'the ringdove' passed by him. She was the mistress of many doves, who were with her. The ringdove perceived the grain, but did not perceive the net, and … they fell into it and were caught in the net together.

Then the hunter came near them quickly, being glad over them; and every dove struggled frantically from her own direction, striving for herself.

And the ringdove said to them: Do not fight with each other as you seek escape, and let not anyone of you be more anxious about her own life than about the life of her companion; but do you all assist each other so that we may perhaps lift up the net, and each of us shall be freed thru the others. They did this and carried off the net, and flew with it into the sky.

The hunter followed them, for he thought that they would go a short distance when the net would become too heavy for them and they would fall.

The crow said: I shall follow them that I may see what is the outcome of this affair of theirs with the hunter.

The ringdove turned around and saw the hunter following them with his hope of them not cut off, and she said to her companions: … direct yourselves to … a place in which I know is the hole of a mouse. He is a faithful friend to me; and, if we go to him, he will cut the net away from us and the injuries we suffer from it.

They directed themselves as the ringdove had indicated, and became concealed from the hunter. And he turned back, having lost hope of them.

But the crow did not turn back, for he desired to see whether they had a trick to employ for extrication from the net, that he might learn it and it might be a resource for him in case this thing should happen to him.

And when the ringdove reached the hole of the mouse with them, she commanded the doves to descend, and they descended, and found around the hole of the mouse a hundred entrances which he had prepared for dangers; for he was experienced and clever.

The ringdove addressed him by name—now his name was Īzāk—and the mouse answered her from his hole saying: Who are you? She said: I am your friend, the ringdove.

He approached her quickly, but when he saw therein the net he said to her: How did you fall into this plight? For you are clever. The ringdove said: … fate has brought me into this plight.… There is nothing strange in my case or my ineffectiveness in opposing fate; for not even he who is stronger and greater than I can oppose fate.…

Then the mouse began to gnaw the meshes in which the ringdove was, but the ringdove said to him: Begin with the meshes of my companions, then come to my meshes.

She repeated the speech to him several times, but the mouse paid

**The crow watches as the mouse gnaws a net to free the ringdoves. Late 16th century. Illustration for *Kalila wa Dimna*. Gujarat, India. Ink and opaque watercolor on paper, 12.5 × 9 inches. The spare artistic interpretation of the passage in the story serves to highlight its essential message. The crow learns the value of friendship from the mouse's actions (Courtesy of Metropolitan Museum of Art, The Alice and Nasli Heeramaneck Collection, Gift of Alice Heeramaneck, 1981).**

no regard to her speech. Then he said to her: You constantly repeat this remark to me, as tho you had no pity for yourself....

The ringdove said: Do not blame me for what I command you, for nothing impels me to this except that I bear the burden of rulership over all these doves, and consequently have a duty toward them. And truly they have paid me my due by obedience and counsel; for thru their obedience and their help Allah saved us from the owner of the net.

But I feared that, if you should begin by cutting my meshes, you would grow weary, and when you had completed that be negligent of doing this with the meshes of some that were left; but I knew that, if you should begin with them and I should be the last, you would not be content, even tho weariness and lassitude should seize you, to avoid the labor of cutting my meshes from me.

The mouse said: This is one of the things that increase the affection and love of those who love you and feel affection for you. Then the mouse began to gnaw the net and continued until he finished it. And the ringdove and her doves went away to their home, returning safely.

When the crow saw the deed of the mouse and the rescue of the doves by him, he desired the friendship of the mouse and he said: I am without safety in a situation like that which befell the doves and I have need of the mouse and his love.

So he approached the mouse's hole. Then he called him by his name, and the mouse answered him: Who are you? He said: I am a crow; affairs have gone so and so with me. I saw your affair with the doves and your faithfulness to your beloved friends, and how Allah benefited the doves thru it, as I saw. I longed for your friendship, and I have come to you for this.

The mouse said: There is no basis for union between me and you. For it behooves the wise man to seek only that which is possible, and to refrain from seeking that which may not be,.... How can there be a way to union between me and you? For I am only food and you the consumer.

The crow said: Consider that my eating you, even tho you are food for me, would not satisfy me in any respect; whereas your continued life and your affection would be more advantageous to me and more conducive to safety as long as I remain alive. You are acting unworthily in sending me away disappointed when I have come seeking your affection. For indeed the beauty of your character has become manifest to me, even tho you do not endeavor to make it manifest yourself.

For the intelligent man—his superiority is not concealed, even tho he strives to conceal it. It is like musk which is hidden and sealed; but this does not prevent its odor from spreading. Do not disguise your character from yourself and do not deny me your love and your kindliness.

The mouse said: The strongest enmity is that of nature, which is of two sorts. The one is an enmity which is equal on both sides, like the enmity of the elephant and the lion, for often the lion kills the elephant, and often the elephant kills the lion; and the other is an enmity in which the injury is from only one of the two upon the other, like the enmity which exists between me and the cat, and like the enmity between me and you. For the enmity with me exists not in consequence of any injury that can come from me to you, but because of what can come from you to me.... But the wise man never associates with a shrewd foe.

The crow said: I have understood what you have said, and you are verifying the excellence of your character. And recognize the truth of my words and do not interpose a difficulty between our relationship by saying 'We have no way to union.' For intelligent and noble men seek union and a way to it for every good purpose.

You are noble and I need your love; and I shall remain at your door without tasting food or drinking until you make friends with me.

...

The mouse said: I accept your friendship, for never in any case have I withheld his necessity from one in need. I began with you as I did merely thru desire of justifying myself, so that, even tho you should be deceiving me, you should not be able to say, 'I found the mouse weak in good sense, easy to trick.'

Then he came out from his hole and stood at the door, and the crow said to him: What keeps you at the door of your hole, and what prevents you from coming out to me and joining me? Have you still doubt?

The mouse said: The people of this world give each other two kinds of things and make alliances on the basis of them. They are the heart and property. Those who exchange hearts are true and loyal friends; but those who exchange property are those who assist and benefit each other that each of them may enjoy the benefit (secured) from the other.... I feel confident in respect to you of your heart, and I present you with the same from me. It is no evil opinion that prevents me from coming out to you; but I realize that you have friends whose nature is like yours, but whose attitude toward me is not like your attitude toward me. I fear that some of them will see me with you and will destroy me.

The crow said: It is one of the marks of a friend that he is a friend to his friend's friend and an enemy to his friend's enemy. I will have no companion or friend who does not love you. For it would be easy for me to cut off from my friendship anyone who is of this sort, just as the sower of sweet basil, when there sprouts among the basil any growth that will injure it and corrupt it, uproots it and uproots some of the basil with it.

Then the mouse came out to the crow, and they shook hands and made friends, and each enjoyed the company of his companion. They remained thus for some days, or as long as Allah wished.

Until when some days had passed for them, the crow said to the mouse: Your hole is near the road of men, and I fear that someone may throw stones at me. But I know a secluded place, and there I have a friend, a tortoise. It is well supplied with fish, and I can find there what I need to eat. I desire to go to her the tortoise and dwell with her in safety.

The mouse said: May I not go with you? For I feel averse to this place of mine. The crow said: Why do you feel averse to your place? The mouse said: I have tales and stories concerning that which I shall tell you when we arrive at the place we have in mind.

The crow seized the tail of the mouse and flew with him until he

> arrived at the place he had in mind. When he drew near the place in which the tortoise was and the tortoise saw the crow and a mouse with him, she was frightened at him, for she did not know that it was her friend, and she dived into the water.
>
> The crow set down the mouse, alighted on a tree, and called the tortoise by name. She recognized his voice, came out to him and welcomed him, and asked him whence he came. The crow told her his story from the time when he had followed the doves, including what had happened thereafter between him and the mouse until they had come to her.
>
> When the tortoise heard of the mouse's deed, she was astonished at his intelligence and faithfulness, and she welcomed him, saying: What drove you to this land?
>
> The crow said to the mouse: Where are the tales and stories which you said you would tell me? Tell them now that the tortoise asks you for them. For the tortoise in her relation to you is in the same position as I. The mouse began his story....

The main story continues along many winding paths, circling around several stories nested within other stories. Toward the end, a fearful gazelle comes to the pond where the three friends are staying. They find that all four must work together to trick the hunter and free the tortoise who has been captured.

On its surface, the principal characters are unnatural friends. In fact, they could be taken for true competitors. As the plot develops, they recognize each other's value and come to realize the benefits of working together. They acquire wisdom from one another and from the adversities they face, and end up creating a strong community. In this world, the animals figure things out and take care of themselves. The human in the story, the hunter, is left behind as an ineffectual being. It's a dreamy "what if" story—what if we could all get along; wouldn't we all benefit?

The crow appears more than once in *Kalila wa Dimna*, one of the most popular books of the ancient world. The history of the work is as fascinating as the many stories within it. Created as a compilation of parables to teach morality, codes of conduct, and the art of statecraft using the actions of animals as metaphors for human behavior, the work is believed to have been written in Sanskrit sometime around the third century BCE, as the *Panchatantra*, by an Indian sage named Bidpai. The collection of stories became so legendary that Iranian Emperor Khusraw I Anushirwan sent a physician named Burzoe to India to retrieve a copy. Burzoe (or maybe a man named Buzurgmihr) translated it into Pahlavi (Middle Persian) in 570 CE. Burzoe's version was itself rendered into Arabic by Abdallah ib al-Muqaffa two hundred years later. This version, considered a masterpiece of Arabic literary prose, became wildly popular. It spawned a vast number of subsequent translations as it traveled the globe.

Kalila and Dimna, two jackals, provide the scaffold for the work. They feature as both protagonists and as narrators of stories. Each chapter begins with the king asking the philosopher about the consequences of certain behaviors and the philosopher uses animal stories to explain. Initial stories spawn additional interior ones as a parade of animal characters come alive through their own tales, discussions, and actions. The animals debate philosophical questions around such topics as ambition, destiny, and the value of wealth. They talk about different kinds of loyalty and how they overcame obstacles, using their own experiences as teaching moments. It is both highly entertaining as well as elucidating.

The stories beg to be illustrated. The anonymous illustration presented on page 144 is one of many Arabic versions to accompany the text. It was painted just after the golden age of Mamluk art (1250–1517 CE) when members of the Egyptian and Syrian court commissioned a vast number of illuminated manuscripts, particularly books with animals. This piece stresses simplicity. It does away with extraneous details to focus on the primary characters and narrative of the story. We see a stylized tree and bold lines. A circle of focus is created by the composition. The curve of the burrow on the right is echoed by the bend of the tree branch above as well as the arc of the net. All noses point inward to the center. The composition, while simple, is not static. The doves' wings flap in all directions to highlight their confinement. The crow is bent so far down in curiosity that it is practically falling off the branch. It is watching and learning. Just like a crow to behave this way....

## *"The Fox and the Crow": Versions of the Aesop Story*

### The Fox and the Crow

By the Rev. Thomas James, 1848.

A Crow had snatched a goodly piece of cheese out of a window, and flew with it into a high tree, intent on enjoying her prize. A Fox spied the dainty morsel, and thus he planned his approaches. "O Crow," said he, "how beautiful are thy wings, how bright thine eye! how graceful thy neck! thy breast is the breast of an eagle! thy claws—I beg pardon—thy talons, are a match for all the beasts of the field! O! that such a bird should be dumb, and want only a voice!" The Crow, pleased with the flattery, and chuckling to think how she would surprise the Fox with her caw, opened her mouth:—down dropped the cheese! which the Fox snapping up, observed, as he walked away, that whatever he had remarked of her beauty, he had said nothing yet of her brains.

*Men seldom flatter without some private end in view; and they who listen to such music may expect to have to pay the piper.*

**Detail of the Bayeux tapestry. Late 11th century. City of Bayeux, France. Embroidered cloth. The fable of the fox and the crow (seen at lower left) is depicted three times on the borders of the Bayeux Tapestry as a political commentary on two men—William, Duke of Normandy and Harold, King of England—who were at the time vying in a high stakes game for power. Some scholars believe William is portrayed as the greedy fox and Harold the foolish crow. Others argue the opposite (Courtesy of Bayeux Tapestry Museum, City of Bayeux, France).**

## The Sycophantic Fox and the Gullible Raven

By Guy Wetmore Carryl, 1899.

A raven sat upon a tree,
And not a word he spoke, for
His beak contained a piece of Brie,
Or, maybe, it was Roquefort:
We'll make it any kind you please—
At all events, it was a cheese.

Beneath the tree's umbrageous limb
A hungry fox sat smiling;
He saw the raven watching him,
And spoke in words beguiling.
*"J'admire,"* said he, *"ton beau plumage."*
(The which was simply persiflage.)

Two things there are, no doubt you know,
To which a fox is used,—
A rooster that is bound to crow,

A crow that's bound to roost,
And whichsoever he espies
He tells the most unblushing lies.

"Sweet fowl," he said, "I understand
You're more than merely natty:
I hear you sing to beat the band
And Adelina Patti.
Pray render with your liquid tongue
A bit from 'Götterdämmerung.'"

This subtle speech was aimed to please
The crow, and it succeeded:
He thought no bird in all the trees
Could sing as well as he did.
In flattery completely doused,
He gave the "Jewel Song" from "Faust."

But gravitation's law, of course,
As Isaac Newton showed it,
Exerted on the cheese its force,
And elsewhere soon bestowed it.
In fact, there is no need to tell
What happened when to earth it fell.

I blush to add that when the bird
Took in the situation
He said one brief, emphatic word,
Unfit for publication.
The fox was greatly startled, but
He only sighed and answered "Tut!"

THE MORAL is: A fox is bound
To be a shameless sinner.
And also: When the cheese comes round
You know it's after dinner.
But (what is only known to few)
The fox is after dinner, too.

primus

¶ Fabula. xv. de Corvo et vulpe.

Qui se laudari gaudẽt verbis subdolis decepti penitent: De quo hec ẽ fabula. ¶ Cum de fenestra corvus caseũ rapet alta consedit in arbore. Vulpes ut hũc vidit, caseum habere cupiens, subdolis verbis sic eũm alloquitur. O corve quis similis tibi, et pennarum tuarum qualis est nitor: qualis esset decor tuus si vocem habuisses clarã, nulla tibi prior avis fuisset. At illa vana laude gaudens: dum placere vult et vocem ostendere, validius clamavit: et ore aperto, oblitus casei ipm deiecit: quem vulpes dolosa celerius rapuit et avidis suis dentibus abrodit. Tũc corvus ingemuit, ac vana laude deceptus penituit: sed post factum quod penitet ¶ Monet aũt hec fabula cũctos verbis subdolis, vaneque laudãtibus nõ attendẽ.

**German translation of Aesop's Fables. 1476. Heinrich Steinhöwel. Germany. Illustration. Around 1436, Johannes Gutenberg brought the printing press to Europe. Aesop's Fables were already immensely popular at the time and were ideal subjects for book publication. The Fables were printed in 150 separate editions between 1465 and 1501. Heinrich Steinhöwel's German translated fables formed the basis for William Caxton's famous 1484 English translation (Bayerische Staatsbibliothek München, Rar. 762, fol. 76r).**

## The Fox and the Crow

By James G. Thurber, 1956.

A crow, perched in a tree with a piece of cheese in his beak, attracted the eye and nose of a fox. "If you can sing as prettily as you sit," said the fox, "then you are the prettiest singer within my scent and sight." The fox had read somewhere, and somewhere, and somewhere else, that praising the voice of a crow with a cheese in his beak would make him drop the cheese and sing. But this is not what happened to this particular crow in this particular case.

"They say you are sly and they say you are crazy," said the crow, having carefully removed the cheese from his beak with the claws of one foot, "but you must be nearsighted as well. Warblers wear gay hats and colored jackets and bright vests, and they are a dollar a hundred. I wear black and I am unique." He began nibbling the cheese, dropping not a single crumb.

"I am sure you are," said the fox, who was neither crazy nor near-sighted, but sly. "I recognize you, now that I look more closely, as the most famed and talented of all birds, and I fain would hear you tell about yourself, but I am hungry and must go."

"Tarry awhile," said the crow quickly, "and share my lunch with me." Whereupon he tossed the cunning fox the lion's share of the cheese, and began to tell about himself. "A ship that sails without a crow's nest sails to doom," he said. "Bars may come and bars may go, but crow bars last forever. I am the pioneer of flight, I am the map maker. Last, but never least, my flight is known to scientists and engineers, geometrists and scholars, as the shortest distance between two points. Any two points," he concluded arrogantly.

"Oh, every two points, I am sure," said the fox. "And thank you for the lion's share of what I know you could not spare." And with this he trotted away into the woods, his appetite appeased, leaving the hungry crow perched forlornly in the tree.

MORAL: 'Twas true in Aesop's time, and LaFontaine's, and now, no one else can praise thee quite so well as thou.

Variations on the Theme

I

A fox, attracted by the scent of something, followed his nose to a tree in which sat a crow with a piece of cheese in his beak. "Oh, cheese," said the fox scornfully. "That's for mice."

The crow removed the cheese with his talons and said, "You always hate the thing you cannot have, as, for instance, grapes."

"Grapes are for the birds," said the fox haughtily. "I am an epicure, a gourmet, and a gastronome."

The embarrassed crow, ashamed to be seen eating mouse food by a great specialist in the art of dining, hastily dropped the cheese. The fox caught it deftly, swallowed it with relish, said "*Merci*," politely, and trotted away.

II

A fox had used all his blandishments in vain, for he could not flatter the crow in the tree and make him drop the cheese he held in his beak. Suddenly, the crow tossed the cheese to the astonished fox. Just then the farmer, from whose kitchen the loot had been stolen, appeared, carrying a rifle, looking for the robber. The fox turned and ran for the woods. "There goes the guilty son of a vixen now!" cried the crow, who, in case you do not happen to know it, can see the glint of sunlight on a gun barrel at a greater distance than anybody.

III

This time the fox, who was determined not to be outfoxed by a crow, stood his ground and did not run when the farmer appeared, carrying a rifle and looking for the robber.

"The teeth marks in this cheese are mine," said the fox, "But the beak marks were made by the true culprit up there in the tree. I submit this cheese in evidence, as Exhibit A, and bid you and the criminal a very good day." Whereupon he lit a cigarette and strolled away.

IV

In the great and ancient tradition, the crow in the tree with the cheese in his beak began singing, and the cheese fell into the fox's lap. "You sing like a shovel," said the fox, with a grin, but the crow pretended not to hear and cried out, "Quick, give me back the cheese! Here comes the farmer with his rifle!"

"Why should I give you back the cheese?" the wily fox demanded.

"Because the farmer has a gun, and I can fly faster than you can run."

So the frightened fox tossed the cheese back to the crow, who ate it, and said, "Dearie me, my eyes are playing tricks on me—or am I playing tricks on you? Which do you think?" But there was no reply, for the fox had slunk away into the woods.

Aesop's fables are among the most widely known animal stories in the Western world. They have been approached from many vantage points. The Rev. Thomas James (1809–1863), a British vicar, took the lessons imparted through the fables seriously, and was keen to present all of them in the most original state possible for the edification of his gentle readers. He said: "It has been the object of the Translator to restore [the work], in more genuine form than has yet been attempted, into the hands of the present generation,…. The general rule has been to give a free translation from the oldest source to which the Fable could be traced, or from its best later form in the dead languages,…."[8] Guy Wetmore Carryl (1873–1904) saw other possibilities. In the 1899 *Fables for the Frivolous*, or *Fables for the Frivolous (with Apologies to La Fontaine)*, Carryl, an American parodist, composed light-hearted poems from what he perceived to be dry material. James Thurber (1894–1961), arguably one of the funniest American essayists of all time, could not resist the temptation to translate Aesop's moral lessons to fit our contemporary times. The results are insightful and hilarious.

To riff on Aesopic material is a long-standing tradition that has continued for millennia. Aesop himself may never have existed; the evidence is skimpy. It has been said that he was a story-telling slave who lived in Greece during the sixth century BCE, at least three hundred years before any of his stories were committed to writing. The fables probably traveled orally through time and varying landscapes to be finally written, collected, augmented, then attributed to Aesop. Even after the first collections and translations were established in his name around the first century CE, new fables were added, and old ones reinterpreted. "The Fox and The Raven" is most frequently attributed to Aphthonius, a scholar and teacher of the fourth century CE. Forty extant fables are attributed to him. The fables have been illustrated in every medium imaginable—in the eleventh-century Bayeux tapestry, as thirteenth century Italian fountain decoration, on eighteenth-century Wedgewood china, as Greek and Hungarian postage stamps. Today, we find the fables transformed into cartoon versions for children (and childlike adults), most memorably as part of *The Rocky and Bullwinkle Show* (1959–1964).

***The Fox and the Crow.*** **Colin Blanchard. 2022. Linocut, 7.8 × 10.6 inches, edition of 6. Parables lend themselves to illustration. Their straightforward simplicity encourages a wide range of visual interpretation (Courtesy of Colin Blanchard).**

Stories designed as animal parables deliver their meaning succinctly. The intent of both *Kalila wa Dimna* and many of *Aesop's Fables* is to convey important messages about how to navigate a complicated and confusing world. They lay out the rules of life in a way we can clearly understand; they show us how to behave (or how not to behave); they counsel us on the pitfalls we will inevitably encounter. And they do so through simple analogy. *Kalila wa Dimna* is long-winded but vivid, using the nesting structure of a story within a story to describe animals acting like humans. In *Aesop's Fables*, animals behave like animals, but sharp lessons are drawn in the spare, open architecture of their parables. We see ourselves in animal behavior. Both works have a graphic quality that renders them totally memorable. In these stories, we witness the crow's natural instincts put to the test in a world of competition and cooperation. In *Kalila wa Dimna*, cooperation wins the day; in Aesop's fable, the thieving crow loses out to someone even more tricky.

# 9

# Sustenance

When reading Aesop's fable about the fox and the crow, we derive just a bit of perverse pleasure to learn that the thieving crow, with a piece of stolen cheese in its beak, ultimately loses out. Corvids often get a bad rap (only partly earned) when it comes to their relationship with our food. We envision them stealing our precious grain and flying off with our baby lambs. They are often described as being pestilential and in need of eradication. During early times, people would station children in their fields to harass the birds away with clackers and pot-banging. In England, feelings ran so high that Henry VII passed the Preservation of Grain Act in 1532 making it compulsory for every person to kill as many crows as possible. A community was punished with fines if it failed to kill enough birds. In eighteenth-century America, states placed a bounty on crow bodies. The practice ramped up in the 1940s and 1950s when the U.S. federal government offered money for the birds in order to protect our nation's grain supply. Residents in Kansas favored dynamiting crow roosts to kill thousands in one blast. Before 1972, crows were shot or poisoned at will.[1] Today, crows are sometimes characterized as comprising marauding death armies that wipe out domesticated and wild animal populations. And this is in some measure true. The incredibly intelligent birds find ways to take advantage of every situation. Scarecrows don't stand a chance at succeeding in their job.

There is, however, another way to look at the place crows and ravens have in our ecosystem, not just as thieves (though they certainly are) but as also beneficial to our lives.

## *Field Notes: A Place Within The Ecosystem*

As scavenging specialists and expert carcass foragers, ravens and crows will eat almost any animal matter (including one another's eggs)—but they favor insects, fruits, seeds, nuts, and grains. This can be a good thing. We have already discussed how their caching habits help the environment by providing food for other animals to discover and by distributing tree and plant seeds across a wide area. The birds also go about delousing sheep and preventing deer from blowfly infestation. They dispense with disease carrying vermin as well as unsightly rotting remains. Crows, in particular, eat insects harmful to field plants. Journalist Jane Brody tells the story of a Midwestern farmer who killed every crow he could find only to discover that the birds had

***Wheat Field with Crows*. 1890. Vincent van Gogh. Auvers-sur-Oise, France. Oil on canvas. 20 × 40.5 inches. In one of Van Gogh's most famous paintings, a flock of crows takes wing between wind-swept fields and a stormy sky. This was a scene Van Gogh commonly saw around his family home. Before the majestic but ever-shifting vista, three paths can be seen in the foreground. Perhaps they are meant to represent metaphorical life choices. One doesn't know which way to turn (Van Gogh Museum, Amsterdam [Vincent van Gogh Foundation]).**

a heartier appetite for the European corn borer than for his corn. Without the work of the crows, his crop utterly failed.[2] The Humane Society is quick to point out, "a crow family can eat 40,000 grubs, caterpillars, armyworms and other insects in one nesting season. That's a lot of insects many gardeners and farmers consider pests."[3]

On the other hand, corvids can certainly be killers. Up north, ravens will kill young seals with repeated blows to the head, and ranchers in the West have witnessed attacks on newborn calves. It is not as common as sometimes imagined (think Hitchcock's *The Birds*) but it happens. Nevertheless, David Rains Wallace reminds us: "They may have tomahawk beaks and four-foot wingspans, but bird bones are fragile and even a sheep has *some* capacity for self-defense. Ravens lack the golden eagle's ability to stun prey with powerful talons and the coyote's ability to attack under cover of darkness."[4]

Crows, too, can be lethal, especially in urban parks, but they mostly cruise our highways, using their excellent eyesight (they see at 100 feet what we see at 10) to identify roadkill and other goodies. Overflowing dumpsters are, of course, favorite haunts and a place where crows and ravens can be picky. They have taste buds on the base of their tongues that enable them to distinguish sweet, bitter, sour, and salty, and they are known to spit out food they don't like. They will also spend time evaluating their food. Researcher Reto Zach studied crow decision-making on the coast of British Columbia. He noticed that crows would fly the beach in search of the best sized whelk—the larger, the better. They dropped these whelks from the precise height that would break them open. If they couldn't find the right size, they moved on to other kinds of food. Zach concluded that in this way they picked the highest calorie food that they could most efficiently access.[5]

Ravens are notoriously cautious around food. Lawrence Kilham replays one scene:

> A coyote killed a lamb on the morning of July 22 but was frightened away before it had a chance to eat. I cut the lamb open, knowing that ravens could not penetrate the hide unless I did so. The family of seven ravens that visited the farm came down but did no more than stand nearby. They did this three times in the course of the morning, without any of them feeding. The carcass remained untouched on the following day, but coyotes ate much of it in the night. The next morning I found nine ravens standing by the carcass, the largest number of ravens I had ever seen at the farm. They approached and jumped back a few times but left without feeding. A little later three ravens alighted at a distance and approached slowly. One started to pull at the carcass. But when a fourth arrived and jumped back, they all left. This happened several times, giving an impression that when one raven jumped, it frightened the others.[6]

Perhaps the fearfulness results from living alongside large predators or maybe it comes from long experience with poisoned carcasses (researchers have suggested both explanations). Regardless, ravens will almost always wait for others to dig into a carcass first. Once they conclude something is safe to eat, ravens "subject potential food to careful taste tests. They delicately crush and palpate new food in their bills for many seconds before either swallowing or rejecting it."[7] When they find something they really like, the birds will go to amusing (and very efficient) lengths to secure it. Bernd Heinrich reports an episode during which an Alaskan oil pipeline worker toyed with a local raven by throwing it two delicious donuts at one time. The raven was unfazed. It stuck its bill through the hole of one donut, then grabbed the second and flew off with both. Heinrich also observed a raven stack four delicate crackers on top of each other and fly off with the stack without crushing them.[8]

The ecosystem in which crows and ravens find themselves operating has been profoundly affected by human activity. We have changed so many features of their world. For instance, prior to 1912 relatively few crow populations roamed Plains states such as Kansas. When homesteaders arrived, they cleared vast amounts of land to grow corn and wheat. By the mid–1920s, crows were coming to Kansas by the millions.[9] Omnivorous habits and the ability to take advantage of a changing landscape have allowed these birds to flourish. When I watch the busy activities of my backyard crows, I make note now of how they navigate "my" territory. I spot them cherry-picking the best worms drying up on my driveway. I also notice them using the highest point on my roof to survey a vernal pool downhill. A few will spread out along the left side of the roofline, heads slightly tipped, to watch the glistening water. Frogs live there. Crows and humans have similar hearing abilities, but birds are more sensitive to the rhythm and tones of sound. They have no external ears so they use their entire heads to detect the location of incoming sound waves. One late afternoon while I was observing them on the roof, my crows suddenly launched themselves into a death spiral toward the vernal pool. They swept through the area picking off dozens of amphibians. I had heard not a single frog peep. I was very, very sorry to lose all those frogs. But I could hardly blame the birds.

## *A Bird Artist: Lilian Marguerite Medland*

In many ways, the life experience of Lilian Medland (1880–1955) mirrors that of most early women artists who chose the natural sciences as their inspiration and career. She existed in the background—as did such brilliant artists as Barbara Regina Dietzsch (1706–1783), Sarah Stone (1760–1844), Elizabeth Gwillim (1763–1807), Graceanna Lewis (1821–1912), and Elizabeth Gould (1884–1841). Their contributions to ornithology and ornithological illustration are barely lauded. Yet Medland was happy in her life.

Born to a wealthy family in Victorian England, Medland carried a passion for illustration but trained as a nurse. She had the good fortune to work with a surgeon, Charles Stonham, who himself was devoted to birds. He asked her to work on the 318 plates required to illustrate five volumes of his book, *The Birds of the British Islands,* completed in 1911. Medland hand-colored the engraved plates. The achievement established her as a competent illustrator, and she was asked to contribute to various other books. The vagaries of war and history, however, prevented five important books on birds containing her illustrations from ever being published.

At the age of 27, Medland contracted diphtheria which left her deaf. A lively, emancipated woman who took up smoking and wore knickerbockers while riding a bicycle, Medland nevertheless continued her artistic pursuits. She married an ornithologist and moved to Australia, eventually painting a series of birds for postcards, illustrating a handbook that depicted 883 Australian birds (which was never published because of World War II), and collaborating with her husband on several significant books, the last of which (on birds of paradise and bower birds) was published in 1950.

Recollections of Lilian Medland describe her as a quiet but curious individual. She was a quick adopter of new technology—taking up photography, installing plumbing, and purchasing the newest gas stove and refrigerator. She was also a keen observer.[10] We see this in the lifelike poses of the birds she illustrated. In *The Carrion Crow*, she created a dynamic scene true to life. The firm, sleek body of the crow is hunched over a limp chick. Its dagger-like beak is poised just above the soft fluff of feathers. The eye of the crow is all clarity. It is ready to begin its gruesome work. We are before nature as a primal force. Yet, the composition of the work is elegant. The curve of the crow's back is echoed in the curve of the surrounding hills and in the body of the chick. The viewer's eye moves easily across the scene. With simple but beautiful directness, Medland was able to educate us to what the carrion crow looks like, what it eats, and the landscape in which it is found.

A treasure trove of unpublished artwork was created by Medland. Her efforts were largely foiled in her lifetime. Today, they are highly sought after by collectors. And, although no book can boast her authorship, she is considered to be one of Australia's finest bird artists.

Both Van Gogh and Medland visually hint at our contradictory feelings about crows and ravens. Nature can feel menacing. These uneasy feelings also show up in the quips and sayings we use. We call a group a "murder" of crows or

***The Carrion Crow.* n.d. Lilian Medland. Illustration in Charles Stonham's *The Birds of the British Islands,* 1905–1911. In his book, Charles Stonham provides a lengthy description of the habits of this maligned bird. Its penchant for eggs, baby poultry, and lambs made it very unpopular among farmers. Lilian Medland gives us a dispassionate illustration.**

an "unkindness" of ravens. To "eat crow" is to admit a humiliating error. Just the sight of a crow or raven flying overhead might be bad luck. Shakespeare's *Hamlet* has us reading: "the croaking raven doth bellow for revenge."[11] In the U.K., "stone the crows" means "I can't believe it." In the Indian Malayalam language, "kakka kulichal kokkakumo" means "a crow cannot become a crane by taking a bath." In Mexico, "Cria cuervos y te sacarán los ojos" means "Raise crows and they will poke your eyes out" (referring to bad parenting style). Even the Bible expresses suspicion. Leviticus 11: 13–15 tells us that they are detestable, unclean animals.

A deeper dive into the mythic past reveals a different point of view. Food is life and corvids are often the ones to deliver this blessing. Origin stories from around the world tell of a crow or raven creating, materializing, or bringing the first food stuffs. In this chapter, we travel the globe to see that in Greece crows are deserving of alms of grain, while in New York State the crow merits a bit of a season's first corn. Among Christians, ravens were believed to have followed God's command in bringing essential provisions to hungry saints. For the Shoshoni, however, life's sustenance arrived accidentally, when Crow dropped his pine nuts all over the place. In Australia, food went from raw to cooked when the crow couldn't put out a fire. Among the Zuni people, food was made edible by the raven. For the Vietnamese, bags of gold were the reward for sharing food with the raven. In other words, these birds can save our lives.

## Rhodian Crow Song

Greece. Translation by Charles Burton Gulick.

Kind friends, give a handful of barley to the Crow, Apollo's daughter;
or a plate of wheat, a loaf of bread maybe, or a farthing-bit, or whatever you please.
Give to the Crow, good sirs, something of what each of you has on hand.
She will accept a lump of rock-salt; yes, she likes very much to feast on that.
Who gives salt now will give honey comb another day.
Boy, push back the door! Abundance has heard us,
and a maid brings figs for the Crow.
Ye gods, may the girl prove to be blameless in every way, and may she find a husband rich and famous;
I hope she may lay a son in the arms of her old father,
and a girl baby on the lap of her mother—
her own offspring to be nurtured as a wife for one of her kinsmen.
As for me, wherever my feet carry me,
I go in turn and sing at the door with tuneful muse,
Whether one gives or does not give more than I ask.
....
Nay, good sirs, hand out some of the wealth which your pantry hoards.
Give, master, and you too, lady bride, give.
It is the custom to give a handful to the Crow when she begs.
That is the refrain I sing.
Give something, and it will be enough.

Imagine a strolling band of singers, accompanied perhaps by a few musicians, as they make their way through narrow streets on the Greek island of Rhodes. They call out for alms—a small donation for the crow. They are known as Coronists and their songs are called Coronismata. Athenaeus, a Greek author living at the end of the second century CE, confided during a dinner-table conversation among philosophers that he learned about the Coronists from the poet Phoenix of Colophon, as well as from the writings of the first-century grammarian Pamphilus of Alexandria and the writer Hagnocles of Rhodes. The practice seems to have been at one time sacred (with the crow acting as go-between among gods and humans) but was adapted to other purposes. To our eyes and ears today it appears a little bit like a cross between Christmas caroling and Halloween trick-or-treat. The singers entertained an audience and asked for something in return.

Wandering minstrels were common in ancient Greece. Rhapsodists, professional itinerant performers of epic poetry, roamed through villages and towns in the fifth and fourth centuries BCE and received donations for their recitals. The *Iliad* and the *Odyssey* were among the favorites in their repertory. Phoenix of Colophon, early third century BCE, may have been a rhapsodist as well. In his work, Phoenix appears as a begging poet, indebted to the Muses but hoping for patronage for his art.

"The Rhodian Crow Song" was apparently sung during wedding ceremonies. Crows were reported to be faithful companions and bearers of important omens, and therefore appropriate birds to invoke. In the song we are asked to feed them (and thus bring good luck). Scholars have pointed out that the sexual references of bringing figs and "pushing open the door" were perceived to be wedding events. This song blessed a marriage and encouraged fertility.

## Saint Paul and the Raven

Retelling from "The Life of Paulus the First Hermit" by St. Jerome.

God sent St. Paul an angel who took him out of this place, and walked with him until they reached the eastern inner wilderness. He stayed there for 70 years, during which he saw no one. He put on a tunic made of palm tree fiber. The Lord sent him a raven every day with a half loaf of bread.

When the Lord wanted to reveal the holiness of St. Paul and his righteousness, He sent His angel to St. Antony (Antonius) the Great, who thought that he was the first to dwell in the wilderness. The angel told St. Antony, "There is a man who lives in the inner wilderness; the world is not worthy of his footsteps. By his prayers, the Lord brings rain and dew to fall on the earth, and brings the flood of the Nile in its due season."

When St. Antony heard this, he rose right away and went to the inner wilderness, a distance of one day's walk. God guided him to the cave of St. Paul. He entered, and they bowed to each other, and sat down to talk about the greatness of the Lord.

In the evening, the raven came bringing a whole loaf of bread. St. Paul said to St. Antony, "Now, I know that you are one of the children of God. For 70 years, the Lord has been sending to me every day, half a loaf of bread, but today, the Lord is sending your food also. Now, go and bring me back in a hurry the tunic that Emperor Constantine had given to Pope Athanasius."

St. Antony went to St. Athanasius, and brought the tunic from him and returned to St. Paul. On his way back, he saw the soul of St. Paul carried by the angels up to heaven. When he arrived to the cave, he found that St. Paul had departed from this world. He kissed him, weeping, and clothed him in the tunic that he asked for, and he took his fiber tunic.

Historically, the raven has come to the aid of several saints—St. Benedict of Nursia shared his food with a raven and in return the raven disposed of a poisoned

*Saints Anthony Abbot and Paul the Hermit.* ca. 1634. Diego Rodríguez de Silva Y Velázquez. Spain. Oil on canvas, 103 × 76 inches. At God's bidding, a raven bullets down to two saints to feed them with bread from his beak. Saint Anthony, at left, wears the black-hooded, brown habit of the Knights Hospitaller. He can be seen in five scenes in this painting which visually recount his story. In the foreground, he appears surprised to see the bread appear. The spare, luminous setting Velázquez chose is reminiscent of the landscape of northern Madrid where he was painting at the time (World History Archive/Alamy Stock Photo).

mouthful meant to kill the saint; ravens avenged the murder of St. Meinrad by tracking down the murderers; St. Oswald and a raven collaborated to convert the daughter of a pagan king; and the prophet Elijah, hiding by the brook Cherith, was sustained by the bread and meat brought by ravens twice a day. St. Kevin is known as the patron saint of crows—he loved black birds so much that he kept a bird nest in his hand until the eggs hatched. The story of St. Paul and St. Antony, however, is perhaps the most widely known of all stories involving saints and crows or ravens. In this legend, the raven, as a servant of God, delivers sustenance to two most holy men. It is included in "The Life of Paulus," written by St. Jerome during his stay in Syria's desert around 374 or 375 CE. The passage has inspired many retellings and just as many illustrations. It is popular because it speaks to the miracle that God can deliver through the agency of something as lowly as a raven and it emphasizes the value of brotherhood and the ascetic life. The mortal men literally break bread together.

St. Anthony is regarded as the founder of the first desert monastic community but, in his biography of Paul, St. Jerome establishes St. Paul as the first hermit. In the third century CE, Paul was driven to the desert of Thebes, Egypt to avoid persecution and there devoted his life to the solitary worship of God until his death at the age of 113 or 114 in 341 CE. He became a model for later hermit saints who were venerated in the Middle Ages and Early Modern times.

## The Theft of Pine Nuts

Shoshoni (Newe), Saline Valley, California.
Collected and translated by Julian H. Steward.

The people in this country had no pine nuts. They talked about going off toward the north to get some.

They started off toward the northeast. Coyote was among them. They went to a big camp where there were many people gathering pine nuts. Soon after they arrived, they began to play the hand game against these people. But the next day they did not know whether they had lost or won. They went on to another place where there were also people who had pine nuts. Here they played a game of shooting at a small round target with a bow and arrow. They bet their lives in this game; the losers were to be killed by the winners. When one side missed the target, its opponents took its arrow. Crow was shooting and had only two arrows left. Coyote watched him. When Coyote saw him losing, he walked around and shouted and wondered what to do. Crow was about to shoot at the target again. Coyote said to him, "Why don't you hit the target?" Crow shot and missed. He had only one arrow now. When he shot this one, he hit the target. Then he began to win. He won back everything they had lost and then won everything the other people owned. Finally, their opponents even bet their pine nuts, and lost them.

The people did not want to give up their pine nuts. They hung them on a tall tree which had no branches, so that no one could climb up. During the night they slept under the tree to prevent anyone getting the pine nuts. Cottontail began to play his flute, "tu hu du du du...." Some old women who were helping to protect the nuts knew that they were going to lose them and began to cry for help. Early in the morning, while the people under the tree were still asleep, Coyote and the others started to get the pine nuts. Coyote said, "What do these old women make a noise for? Why don't they go to sleep?" He poked their eyes with a stick and blinded them. Woodpecker (a red woodpecker) flew up in the tree and took the pine nuts.

When Woodpecker brought the pine nuts down, Coyote and the people took them and began to run for home. The others pursued them and caught those who became tired while they were running. They killed everyone they caught. Although many people started out, nearly all were killed before they got home.

When nearly all the people were dead, Woodpecker gave the pine nuts to Crow. Crow went on with them. He hid them under his feathers, behind his ear, and in other parts of his body. The pursuers knew he would hide them this way and tried to hit him. They struck his leg and knocked it off. It went a long way through the air. Then they struck Crow and brought him to the ground. They said, "Now we will wait and take a rest."

After they had rested, they went on to where Crow had fallen and searched his body for pine nuts. They found that Crow had left his feathers behind [i.e., shed his skin] and had gone on, taking the pine nuts with him. They looked and a long way off saw where his leg had fallen, but Crow was far beyond, still carrying his pine nuts. They saw pine nut trees all over the mountains where the nuts had fallen from Crow's leg when it was knocked through the air. They saw smoke coming up through the trees, where the people were already picking the pine nuts. Crow was flying about crying, "Caw, caw, I have had my pine nuts with me all the time." All this happened up by Lida.

A vast expanse, encompassing dry desert, tall mountains, deep valleys, lakes, and rivers, offered the Shoshoni people (or *Newe*, "The People") abundant plants and wildlife from which to live. Their territory stretched from southern Idaho through parts of northwest Utah, Nevada, and into the Death Valley region of California. Their mythology centers on a monumental trickster figure, Coyote, who is a lustful, mischievous, playful culture hero, but also a very powerful and somewhat dangerous being.

In Shoshoni mythology, the supernatural world seamlessly weaves through the human and natural world. Mythic animals can appear as natural animals or act like people. They desire the same things and will compete, cooperate, kill, and help one another to get what they want. In "The Theft of Pine Nuts," a Western Shoshoni myth, we read about a competition for one of nature's precious resources, the extremely nutritious pine nut (high in protein, iron, vitamins, and unsaturated fats). Crow defends his possession of the pine nuts but inadvertently distributes them across the landscape. This was a helpful outcome and something that has since inextricably tied the supernatural, natural, and human together.

Julian Steward (1902–1972) spent time among the Shoshoni, Paiute and other Indigenous groups of the Great Basin and Plateau regions in order to examine his theory about the importance of environment on the development of societies. He rejected popular theories of the time which held that culture and history should be used exclusively to explain the unique composition of a society, and instead created an ecological framework, which he coined "cultural ecology," for describing cultural progress through time. Steward privileged environment over any other factor affecting a people (minimizing the influence of religion, social organization, internal political institutions, external colonial forces, or population dynamics). He was criticized for having an excessively narrow focus as well as for some of his methodologies, but Steward is credited with shedding new light on the environmental factors that affect lifeways.

## Why Crow Has the Right to the First Corn

Iroquois (Haudenosaunee), New York. Collected and translated by Harriet M. Converse.

Among the birds which came from the sun land, *Ga-gaah*, the Crow, carried in his ear a grain of corn which *Hah-gweh-di-yu* (Spirit of Good) planted above the body of his Mother (the earth), and it became the first grain, the "life" of the red man. By this birthright, *Ga-gaah,* claiming his share, hovers above the fields, guarding the young roots from the foes which infest them.

The Iroquois confederacy is made up of six nations—the Mohawk, Cayuga, Oneida, Onondaga, Seneca, and later the Tuscarora. When encountered by white settlers, they were formidable peoples who occupied vast stretches of resource-laden land from Quebec to the southern Great Lakes and New York state regions. They referred to themselves as *Haudenosaunee,* or "people of the Longhouse (their multi-family dwellings)." Early colonists recognized the immense natural wealth to be found in their deep forests, fast-moving rivers, wide lakes, and arable soils, and immediately engaged them in trade and land dealings that ultimately led to total land divestment, near genocide, and cultural loss. Not erased, however, were their stories.

Much of Iroquois society was dependent on corn, of which there were five varieties. We learn here that corn came by way of a crow from the sky (or sun) world. It was believed that after Crow's introduction of corn to the land, *O-na-tah* (Spirit of the corn and patroness of the fields), would initiate the aid of the sun when she inaugurated each planting season. O-na-tah's crow flocks would whirl and call in the sky then swoop down to hunt the creatures who burrowed into the earth to eat the new roots. (The Iroquois, nevertheless, defended their growing corn against crows by placing platforms and sentinels in the fields, setting snares, even hanging dead or living crows from tree branches.) Crows and ravens were among the most magical of "medicine" creatures in the Iroquois universe, and also the wisest, because they had their own chiefs and councils.[12]

Stories filled the long winter nights in the snow-bound longhouses. Each community had its own official storyteller who had been taught all the legends of the past. The storyteller would announce his intention to begin a tale and expect an eager reply of "Hēh!" from those assembled. At intervals, exclamations of "Hāh!" were thrown out to indicate that all listeners were still with him. No one was allowed to fall asleep without permission. Although all the stories were well known, each narrator was proud to claim his own style and version. Folklorist Arthur C. Parker wrote:

> The Iroquois were a people who loved to weave language in fine metaphor and delicate allusion and possessed a language singularly adapted for this purpose. They were unconscious poets, and some of their tales seem to have been chanted in blank verse, the rhythm and swing of the meter in their estimation giving an added delight to the story. When the legends are told to white men the delicate word-weave is seldom revealed, and never if the legend is told in English.[13]

Harriet M. Converse (1836–1903) was able to hear many of their stories. This independently wealthy poet, living among New York's elite echelon, embraced the Iroquois and focused her life's work on collecting their myths and cultural artifacts, recording their history and traditions, and advocating for their rights. She supported them by establishing boarding houses for them in New York City and defending their territorial claims in court. In recognition of her contributions, she was made a member of the tribe and given an honorary position of chief. What is not usually reflected upon, however, is the effect of her role as "chief spokesperson" for the Iroquois. Converse was their default intermediary and negotiator. She gobbled up their possessions for numerous museums and framed Western perceptions of them even as she defended their interests. Like other nineteenth century fieldworkers, Converse involved herself in "salvage ethnography," a kind of predatory capture of a supposed dying culture's patrimony, even as she deeply desired to help them.

## THE FIRST CORN

Zuni (A:Shiwi), New Mexico. Collected and translated by Ruth Bunzel.

When all their days were passed, gathering together their sacred possessions, and arising, hither they came. To the place called since the first beginning Upuilima they came. When they came there, setting down their sacred possessions in a row, they stayed quietly. There they strove to outdo one another. There they planted all their seeds. There they watched one another's days for rain. For *k'ä-eto:we*, four days with heavy rain caressing the earth. There their corn matured. It was not palatable, it was bitter. Then the two said, "Now by whose will will our corn become fit to eat?" Thus they said. They summoned raven. He came and pecked at their corn, and it became good to eat. "It is fortunate that you have come." With this then, they lived.

**Raven, Zuni fetish. 2022. Brian Yatsattie, Zuni Pueblo. Carved Siberian jet, approx. 4.5 × 1.25 × 2.5 inches. A fetish protects and helps its owner. It must be cared for. Fetishes can be made of any material and can be owned by an individual, family, clan, kiva society or by the entire tribe. The raven is thought to help in healing (image courtesy of Keshi: The Zuni Connection, Santa Fe, NM).**

During a "talk concerning the first beginning," a Zuni man described for Ruth Bunzel the history of his people's origin. It is a long, lovely story of creation and wandering. Pieces of the world were created as the people repeatedly gathered themselves and their sacred possessions together to travel from one place to another. They finally settled in the Middle Place or *Halona: Itiwana*, Zuni Pueblo. There they have lived since the beginning—in multi-story adobe homes surrounded by the wide-open spaces of western New Mexico.

The Zuni, who call themselves *A:Shiwi*, developed an intricately complex religious and ceremonial existence that integrated all the forces around them. Bunzel explains: "To the Zuñi the whole world appears animate. Not only are night and day, wind, clouds, and trees possessed of personality, but even articles of human manufacture, such as houses, pots, and clothing, are alive and sentient. All matter has its inseparable spiritual essence." She adds: "Man is not lord of the universe. The forests and fields have not been given him to despoil. He is equal in the world with the rabbit and the deer and the young corn plant. They must be approached circumspectly if they are to be persuaded to lay down their lives for man's pleasure or necessity."[14] Ritual and ceremony were ways to approach other beings and to engage with them for the benefit of all. We read here about the role that a raven played in making corn edible to humans. In return, even a dead bird would receive a tiny offering of cornmeal. But the Zuni also created fetishes, songs, costumes, dances, and prayers with which to communicate across boundaries. Their world was teeming with interspecies communication.

When Ruth Bunzel (1898–1990) arrived at Zuni Pueblo in the early 1920s, the village had already encountered a number of anthropologists interested in their culture. Frank Cushing (Chapter 2) had been among the first, and he was able to live among them freely. For the Zuni, however, that experience as well as subsequent ones left them wary and defensive. They were offended by the cavalier way their sacred information and objects were collected and disseminated. In most cases, permission was neither requested nor given. Bunzel had difficulty finding a place to stay as she was forbidden to live with any family. Her attitude of empathy and self-awareness as well as her passion for their art allowed Bunzel to make some friends, and she returned to conduct research on their poetry, myths, and ceremonies.

The Zuni were able to sustain their rich heritage. Within a landscape of valleys and mud flats, mountains, mesas, and arroyos filled with yucca, sage, and cacti, the traditions live on today, to be handed forward to new generations.

## How the Kamilaroi Acquired Fire

Kamilaroi, North Central New South Wales, Australia. Collected and translated by R.H. Mathews.

At one time the crow was the only one acquainted with fire and its uses. When the other people had been eating game, blood was always observed around their mouths and jaws, but nothing of that kind was ever noticed about the crow's face. Being questioned on the subject, he said he always cut his meat into small pieces with his stone knife, but his answer was not considered satisfactory. He was invited to a corroboree [party] where some comical fellows were to perform. After a number of clever dancers had taken their turn, without disturbing the crow's equanimity, the shingle-back [type of lizard] and sleepy-lizard danced along by the camp-fires, singing:

*Yamburngain bumbaingo nyi dhu-u-ra*
*Gunaga bid-yeringga būmbūl guna-guna.*
*(content not suitable for translation)*

All the time they were performing, the ordure was trickling down their legs, and when they gave a special jump there was an extra discharge of it. This so completely engrossed the crow's attention that the sparrow-hawk, Gur-gur, came up beside him, catching hold of a little bag containing fire, and running away with it. When the crow saw what had happened, he rushed after Gur-gur, and in the scuffle the fire got jerked out of the bag, speedily igniting the dry grass and leaves. The crow tried his best to prevent the fire from getting away by stamping upon it with his feet, and when that did not succeed he lay down full length and rolled over and over among the burning grass, but all his attempts to recover possession of the fire were unavailing. It spread through the whole country, so that all the people had their share of it, and have used it ever since for cooking and other purposes.

Even a veteran trickster can be fooled if he is distracted in just the right way. This time a sparrow-hawk was able to steal the bag of fire from out of Crow's talons. Crow lost control of the precious resource, but all the people benefited. They could finally cook their food instead of eating it raw. The story of Crow and fire takes a different turn among the Wurundieri people of the Kulin nation (south central Victoria). In that popular story, Crow is the one to trick seven women into giving up their live coals so that he can cook his food. In both cases, the fire gets out of hand, burning Crow as it spreads to mankind.

As mentioned in Chapter 1, Crow plays a central role in Australian Aboriginal culture and mythology, featuring in many stories of the Dreaming (a time without beginning or end). For the Kamilaroi nation, as for many others, the bird can be integral to identity. At birth, each child is associated with a totem that links them to the Dreaming, the living world, and their ancestors. If the crow is your personal totem, the two of you maintain a special relationship, one that is watchful and protective. The crow will send you important messages. Crow also commands the allegiance of one of the two clan groups of the Kamilaroi. (Eagle is the other.)

But Crow is probably not really a crow. In Australia, there are three native types of raven and two types of crow. They can hardly be told apart. The Australian raven, a large bird measuring 18–21 inches, is the most common in the south where the Kamilaroi have lived for close to 40,000 years in an environment built around river resources.

By the late 1800s, when this story was collected, the Kamilaroi population was already in tremendous decline. Robert Hamilton Mathews (1841–1918), whose family had fled Ireland in 1839 amid a tornado of penury and murder charges, encountered them in the course of his work as a successful land surveyor. Mathews was enthralled with Aboriginal culture and began, in his early fifties, to collect information about their ways of life. He had no formal academic training for this work but was an assiduous, self-taught student of anthropology who had the benefit of friendly ties within Indigenous communities. He stayed with the native peoples but also used farmers, mission managers, and policemen to collect information. Mathews was maligned during his life by defensive professionals working in Australia but was lauded abroad by anthropologists and social scientists. Today his work is looked upon as an important resource in spite of the fact that he rephrased narratives to fit Western norms, censored parts he deemed obscene, and wrote little about the circumstances in which the stories were told (location, song, and performance were often critical to the execution of a story and these elements were left out). His publications remain some of the earliest records made of the peoples who were the first inhabitants of Australia. They are revisited again and again by First Nations people as well as outsiders.

## The Raven and the Star Fruit Tree

Vietnam. Retold by Ellen Williams.

A long time ago, a wealthy man lived in a small village with his two sons. The boys grew up to be very different people. When the man died, he left both of them his fortune. Right away, the elder son claimed the whole thing, allowing his brother to have only a star fruit tree, a *cay khe*. The younger brother was a kind and good soul, and he took care of his *cay khe*. He tended to its branches and watered its roots. He hoped that it would produce fruit that he could sell at the market. His brother, on the other hand, went around the village with nothing to worry about.

When the *cay khe* fruit was nicely ripe, a raven, a *con qua*, flew by and stopped in the branches. He tasted the delicious fruit. The younger brother was dismayed to see the bird eating up the fruit he was going to sell, but didn't know what to do. Finally, he decided to speak to the *con qua*.

"*Con qua*, this *cay khe* tree is all I have. If you eat all the fruit from it, my family will starve. Please, please don't."

"Don't worry, my friend," the *con qua* answered. "I will pay you back in gold. Make yourself a bag two feet long to hold the gold."

The brother trusted these words and ran to his wife to make a sturdy bag for him. The next day, the raven returned as promised. He told the man to sit on his shoulders, between his wings. They took off into the air and flapped toward a distant island of gold. There, the man filled his bag.

After they got back, the man and his wife were able to live a comfortable life.

One day, the man decided to invite his brother to a family meal. The elder refused. He did not want to visit any poor relative. But after much pleading, he agreed to come. When his family arrived to the house, he was surprised to see how much had changed. All evening he pestered his younger brother about how he came to such good fortune. The younger brother told him about the raven and the star fruit tree. After he heard the story, the elder brother begged to trade all his fortune for this tree. Out of respect for his elder brother, the younger one agreed to take the family house and land in exchange for the tree.

***The Raven and the Star Fruit Tree*. 2023. Shin-Yeon Moon. Digital illustration. A lesson about family, trust, respect, and greed is told in this beloved Vietnamese story (©Shin-Yeon Moon).**

When the tree blossomed the next season and the *con qua* arrived to eat the fruit, the elder brother insisted that he be taken to the island of gold. The raven agreed. The next day the *con qua* landed on the ground to take the man on his shoulders and off they flew. But the brother was so greedy that he had made a bag twice as big.

They departed for home, but the load was so heavy that the raven finally said "Drop the bag, drop the bag! I cannot carry all the weight!"

The elder brother refused, replying "I will never allow my brother to be richer than I am."

The tired raven could not hold the man and his gold. He dropped the elder brother into the sea.

The brother's family waited and waited for his return until a new season began and the raven returned to eat the star fruit. The younger brother asked what had happened. The raven explained "Your brother has all the gold he wanted."

In Vietnam's crowded cities, in its jungle villages, and around its rice field outposts, the fable of "The Raven and the Star Fruit Tree" has been told countless times. It has been pictured in print and animated film. In 2022, it was the subject of a series of the nation's postage stamps. The original version of the tale is lost to history, but its lessons still resonate with meaning. This most beloved of Vietnamese stories instructs us to beware the dangers of avarice and family rivalry. It guides us to be good to our siblings and not to be too greedy. The *con qua* (raven) plays a neutral but pivotal role. The bird rewards the one who allows it to eat from the precious star fruit tree. It is magical enough to know where the land of gold lies and big enough to carry a human there. In the story, the interdependence of man and nature is neatly encapsulated—in an unselfish world, humans, animals, and plant life can live and thrive, but disaster results when greed tips the balance.

Vietnam's vivid folk telling tradition blossomed in its village festivals, which combined religious ritual, folklore, puppetry, music, and dance. Several types of stories were told on these occasions: fables that commemorated village heroes, fairy tales with magical qualities, humorous stories about everyday people, and more serious allegories used to impart moral or ethical standards. A blend of influences, from Indigenous beliefs to Buddhism, Taoism and Confucianism, can be found within all the stories. They are notable for their elegance, gentle tone, and emphasis on family values.

A mystery does circle around the fact that Vietnam has no native ravens. Why are they consistently identified as such? The only plausible guess must have to do with translation. Vietnam has four corvid species. The large-billed, or jungle, crow looks very much like a raven. It is not much larger than an American crow but has a square head and a long, thick, dagger-like bill that is curved on top. An outsider must have misidentified this con qua (crow) as a raven and the appellation stuck.

# 10

# Transformation

To become something new and different—it's a possibility that ignites the imagination. What would happen if we took another, more alien, form? What would we do? How would we feel? People today routinely project human qualities onto animals and, conversely, give animal attributes to humans. A man can be "dogged" in his efforts; a child can "chicken" out in the face of a challenge. In early times, borders were passable and shape-shifting was common. Knud Rasmussen recalled an Inuit woman tell him: "In the very first times ... both people and animals lived on the earth, but there was no difference between them. They lived promiscuously: A person could become an animal and an animal could become a human being. There were wolves, bears, and foxes but as soon as they turned into humans they were all the same."[1] Traditional shamans carried forward this special ability to morph into animals. They did so to help their community.

In places such as Europe, we encounter ancient stories of people locked into an animal form. In *The Odyssey*, the goddess Circe turned Odysseus' sailors into pigs. In England, King Arthur was thought not to have died but to have lived on as a raven. Sometimes, transformation was a form of punishment. The Greek god Artemis turned Calisto into a bear because she broke her vow of celibacy. Athena transformed Arachne into a spider for challenging her weaving skills. In a Slovak story "About the Twelve Ravens," a mother turns her sons into ravens because of their disobedience and gluttony.[2] Today, we are more likely to fantasize about mermaids and werewolves moving among us. And some of us describe ourselves as "therians"—individuals who believe they are non-human animals (in a non-biological sense). Therians say they experience "species dysphoria." They feel more aligned with an animal self than a human self.

Transformation can a potent, liberating condition. It offers us the chance to be more than we are—to gain a new awareness, obtain power, or build connections. Who wouldn't want to know what it's like to be an eagle, a whale, or a frog? Ravens and crows are notorious avatars; as mythological figures, they quite naturally take new forms when it suits their purposes. Which is why we are so drawn to them. We know how easily they move across worlds. In this chapter, we explore works that put color, shape, detail, and meaning to the experience of becoming something other than yourself (whether that be human or bird). But first, a little bit about a bird's natural transformation....

## *Field Notes: From Newly Hatched to Adult Bird*

A sensational kind of metamorphosis occurs every year in nature when any newborn grows into an adult. It is miraculous to witness an animal grow so quickly from helpless creature to capable member of its species. The transformation of a corvid hatchling into a full-grown bird is just as amazing. At birth, a young raven is blind and naked, with a huge wobbly head and soft bones. It cannot maintain its body temperature and will die from the cold if not for the warmth of its mother. Within eight days, the bird will have increased its body weight by about twelve times. In just six weeks, it reaches adult weight. This rapid growth takes a lot of food! Raven parents will make as many as fifty food visits each day to feed the young who never stop yelling for more. The same pretty much holds true for baby crows whose piteous cries can sometimes be heard from the ground.

In contradiction to their reputation for abandoning their young, raven parents are in reality devoted to their chicks. Father and mother work together. At first, the male will feed the whole family; after about two weeks, when the chicks have some

***Above and opposite:*** **American crow. Weeks 1 and 4. The development of a corvid from newly hatched to mature is surprisingly fast. But even though they may be full grown in a few weeks, they still have a lot to learn to be adult members of their species (Photograph #1 courtesy of Grace McDermith, Santa Barbara Wildlife Care Network; Photograph #2 courtesy of Maureen Eiger, director at Help Wild Birds, Roanoke, VA).**

body cover, the female will join in the procurement of food while the male watches out for enemies. Naturalist Fred Bruemmer writes: "Raven families stay together, play together and feed together all through the summer. The youngsters learn from their parents how to find food and how to recognize enemies. They also acquire some of their elders' wisdom. On their own, young ravens can be pathetically naive."[3] Like other birds, only about half the chicks survive their first year. Starvation, adverse weather, and attacks by larger birds or mammals (such as cats, raccoons, coyotes, and humans) take their toll.

A species as intelligent as the corvid engages in a complex social life that requires a long time to learn. It can take two to three years to acquire the necessary skills for independence. Competence derives from family lessons as well as life experience. During the early days, older juveniles within a crow family unit will act as family helpers. They will assist in feeding, nest repair, cleaning, and guard duty. This setup benefits both the parents and the helpers who continue to mature in the cooperative group setting. The young birds are guided by those who stay close at hand to warn about dangers and to demonstrate techniques of flight and foraging. Corvids, ravens in particular, cleave to social hierarchy. If a youngster fails to understand shifting dominance relations, chaos will ensue. Bernd Heinrich writes: "It would be difficult to overestimate the importance of learning in the life of ravens … as every parent knows, an important aspect of the learning process is simply gaining exposure to what is important."[4]

A young crow spends about a month in the nest before taking its initial flight. This is a time of high risk and mortality for both crows and ravens. Naturalist Lyanda Haupt describes the moment:

> For days before they actually jump, you can see the small birds readying themselves. They cling to the edge of the nest with outstretched wings, teetering as they practice flapping. Sometimes the nestlings, step cautiously out of the nest, line themselves up along a branch and sidestep back and forth along it for a day or two, gathering courage before taking the big leap. Twice I have seen an adult encourage a ready-but-reluctant chick out of the nest and onto the branch by placing food a foot or so beyond the young bird's reach. When the young crows finally do jump, they are often not at all ready for flight. They flap wildly as they crash to earth, where they crouch wide-eyed over their fat bellies, dazed and astonished.[5]

Jane Rudick of Montgomery, Alabama, told writer Michael Westerfield a wonderful story about what happened to a baby bird she found walking about her yard and named Nevermore: "About five crows were standing guard in the trees cawing enough to wake the dead. The baby was fully feathered so we thought it could fly and put it up on the roof, but it just walked around 'til it fell into the back yard." Rudick took it into her house and fed it for three weeks, letting it out in the mornings. All the while, four guardian crows came down to see it twice a day. When it could fly a bit, nine crows came to visit. She wrote in her journal: "If I do not have him outside by 5 a.m. the crows start up a racket outside my door waiting for him. It is remarkable that not a day has gone by that they have not come and cared for him. They have never given up." Just about a week later she let the bird out for the whole day.

> There were more crows, about twelve this time, and I went inside while they swooped down, perhaps showing him how to fly. They watched from the Magnolia trees and telephone poles and four landed, then eight, and they hopped right up to him and did a little dance with their feet, and fanned their wings, and made a lot of excited noise. After thirty minutes all of them left except three. Sunday at noon Nevermore wanted out and was fanning his wings so I decided that I had time to try again. I cawed and the family which was used to my routine was not close by but managed to come within a few minutes to watch and cheer. Nevermore took off and made it up to the high branch of a tree.... I held [the neighbor's] dog at the fence for an eternity till Nevermore flew to a pine tree with his family.... It is a joy to know that Nevermore has graduated to the tall trees in the wood. I was truly amazed that his family remained every loyal to him and continued to encourage him throughout the many weeks.[6]

So the transformation of a crow or raven is both physical and mental. Like humans, they become wholly different creatures from what they started out as. And like humans, corvids mate for life. A raven pair will use the same nest for years; a young crow will return to its home nest if its mate dies or its own nest is unsuccessful. Eventually all young birds leave their homes for other locations to establish their own families, but they don't appear to forget their roots.

## *A Bird Artist: Leonard Baskin*

Leonard Baskin (1922–2000) was compelled by the mythical possibilities of transformation. The image of a helpless baby chick growing into an ordinary bird did not appeal to the renowned artist. Baskin wanted to know: What would be unlocked if we combined two species? What crazy combustion would happen? He drew and painted many crows and crow-men.

In *The Crow* we see a hybrid birdman that is truly disquieting. He stands like a human, on two sharply clawed legs. His humped shoulders seem to shrug as he turns his beaked head away from us. It's as though he's trying not to notice us. Birds do that. The clothing appears feather-like, shredded, molting. Is this a bird changing into a man, a man morphing into a bird, or maybe an individual trapped in between two states of existence? The image shows strong, expressive strokes which give it both a sculptural and a transparent quality at the same time. The birdman stands alone on a blank field. His hollowed-out form seems to speak to a despair of some kind.

Indeed, Baskin was drawn to the anguish of our human condition. He called himself death-involved and crow-haunted. His work constantly circled around vulnerability, transitoriness, and tragedy. Baskin drew inspiration from ancient mythologies where supernatural and human entities fought epic forces, seeing in them universal themes still relevant to the modern age. He had a mytho-poetic way of thinking, which made him an excellent partner for the poet Ted Hughes. They worked together in long collaboration and creative partnership. Their work resulted in the publication of one of the most lauded poetry books of the twentieth century, *Crow* (1972), a modern epic rooted in the Trickster tradition.

**The Crow. n.d. Leonard Baskin. Pen and brush and black ink, with brush and gray wash, over traces of graphite, on ivory watercolor paper, 31.31 × 22.50 inches. In the hands of Leonard Baskin, "beast" and "man" merge. We see here something that is the antithesis of a classic angel (the woman with wings of white) (Adelaide C. Brown Fund, Art Institute of Chicago; ©Estate of Leonard Baskin).**

Baskin was particularly sensitive to the magic that happens when word and image combine. He understood how the graphicality of language mirrors the narrative bent of the visual. While he grew up in Brooklyn the son of a Lithuanian Orthodox rabbi and was steeped in the Jewish religion, Baskin knew he wanted to be an artist as a teenager. He attended New York University as well as Yale University. While still at Yale, he founded his own publication for art and literature, The Gehenna Press, one of America's first limited edition fine art presses. The Gehenna Press gained national recognition for its sophisticated, elegant typesetting and for the excellence of its workmanship. It set a new example of the book as an art form. The Press published over 200 books in sixty years.

Baskin's own expressions of raw intensity extended to a variety of art forms. He was a sculptor (in wood, stone, and bronze), a painter, an illustrator. When other artists of the 1950s and 1960s were abandoning the human figure to abstraction, Baskin made it pivotal to his work. He grappled directly with the grotesque nature of our existence, as if facing it head-on would offer a way through to some kind of uplifting resolution. The crow became a kind of metaphor for humanity. He saw in both species (human and bird) the will to survive—by force or by wit, through ridiculousness or deadly seriousness, and, if need be, by way of tricks and thievery.

## A-Ni'äne'thahi'nani'na Nisa'na

Arapaho (Hinono-eino), Western USA. Composed by Sitting Bull; translated by James Mooney.

My father did not recognize me (at first),
My father did not recognize me (at first),
When again he saw me,
When again he saw me,
He said, "You are the offspring of a crow."
He said, "You are the offspring of a crow."

Transformation was the hope of the members of the Ghost Dance religion (described in Chapter 5). The nineteenth century was a bad time for native peoples across the country. Their land had been stripped from them, they were suddenly unable to provide for themselves, they died of foreign diseases, and they were murdered by every kind of white man. Relief could only be found through the rituals, dances, and songs of a religion which would literally transform the world. It was believed that the foreigner would disappear, the buffalo would reappear, and their deceased loved ones would descend to earth to reunite with them. As we have seen, the crow played an active role in this renewal.

"A-Ni'äne'thahi'nani'na Nisa'na" was composed by Sitting Bull, who is identified by anthropologist James Mooney as an apostle of the Ghost Dance religion. Sitting Bull here relates the experience (while in a trance) of meeting his father who had died years before. The expression, "You are the offspring of a crow" refers to Sitting Bull's sacred attributes, the crow being a messenger to and from the spirit world. The

**Ghost dance shirt. ca. 1890. Hinono'ei, Southern Arapaho. Oklahoma. Tanned elk hide, magpie feathers, pigment, feathers. Shirts were carefully made to convey the messages of a vision encounter. They were often believed to stop bullets when worn. This shirt displays symbols of the spirit world—blue sky and stars, crows and magpies surging toward a red figure with outstretched arms (Courtesy of the Buffalo Bill Center of the West, Cody, Wyoming; Plains Indian Museum; Chandler-Pohrt Collection, Gift of the Searle Family Trust and The Paul Stock foundation; NA.204.5).**

crow had the ability to cross into the other world, join the community of the deceased, and guide them back to a renewed earth of peace and prosperity. By singing this song, Sitting Bull signaled that he was a leader whose authority drew from the spiritual as well as the political.

Sitting Bull (ca. 1831–1890) was widely acknowledged as a holy man of great wisdom, and the brave warrior and chief of the Lakota Sioux nation. During the mid–1800s, the Sioux, Cheyenne and Arapaho were engaged in a determined resistance against United States government policies to remove tribal peoples from their lands. Sitting Bull was an inspiration and role model for this resistance. He was involved in many battles, eventually held as a prisoner of war at Fort Randall in South Dakota, then allowed to live in Standing Rock Reservation. His association with the rise of the Ghost Dance, however, led to new government tensions. Indian agents feared that the Lakota leader would flee the reservation with the Ghost Dancers and therefore ordered his arrest. He was killed in a resulting skirmish during which more than a dozen people died.

Sitting Bull's song resonates today. The turns of phrase (from "my father did not recognize me" to "he saw me" to "you are the offspring of a crow") speak clearly to the confusion of the moment they were all in and to the hope they still had.

## Raven Meets Man

Eskimo (Inuit), Bering Strait. Collected and translated by Edward W. Nelson.

It was in the time when there were no people on the earth plain. During four days the first man lay coiled up in the pod of a beach-pea. On the fifth day he stretched out his feet and burst the pod, falling to the ground, where he stood up, a full-grown man. He looked about him, and then moved his hands and arms, his neck and legs, and examined himself curiously. Looking back, he saw the pod from which he had fallen, still hanging to the vine, with a hole in the lower end, out of which he had dropped. Then he looked about him again and saw that he was getting farther away from his starting place, and that the ground moved up and down under his feet and seemed very soft. After a while he had an unpleasant feeling in his stomach, and he stooped down to take some water into his mouth from a small pool at his feet. The water ran down into his stomach and he felt better. When he looked up again he saw approaching, with a waving motion, a dark object which came on until just in front of him, when it stopped, and, standing on the ground, looked at him. This was a raven, and as soon as it stopped, it raised one of its wings, pushed up its beak, like a mask, to the top of its head, and changed at once into a man. Before he raised his mask Raven had stared at the man, and after it was raised he stared more than ever, moving about from side to side to obtain a better view. At last he said: "What are you? Whence did you come? I have never seen anything like you." Then Raven looked at Man, and was still more surprised to find that this strange new being was so much like himself in shape.

Then he told Man to walk away a few steps, and in astonishment exclaimed again: "Whence did you come? I have never seen anything like you before." To this Man replied: "I came from the pea-pod." And he pointed to the plant from which he came. "Ah!" exclaimed Raven, "I made that vine, but did not know that anything like you would ever come from it. Come with me to the high ground over there, this ground I made later, and it is still soft and thin, but it is thicker and harder there."

"Raven Meets Man" is the first part of a very long, utterly charming, origin story of first encounters, obtained by Edward W. Nelson during a residence in Northern Alaska at the fur-trading station of St. Michael, 65 miles north of the Yukon delta and 200 miles south of Bering Strait. The story was recited by an old Unalit man living at Kigiktauik, who learned it, when he was a boy, from an old man. The narrator said that the individual from whom he learned the story came from Bering Strait, and that always, when he finished his tales on the third evening, he would pour a cup of water on the floor and say: "Drink well, spirits of those of whom I have told."[7]

Here we catch Raven at the moment he sees the first man. He does not recognize the creature, yet we are led to believe they are the same kind of being. After all, Raven "raised one of its wings, pushed up its beak, like a mask, to the top of its head, and changed at once into a man." The transformation was instantaneous. Later on in the story, Raven draws down the mask over his face, changing again into a bird, to fly into the sky and obtain food for the man. He changes back and forth as needed to get the first man everything to live by. In this way, humans were able to populate the earth and thrive. It must have been great to be Raven. He could go anywhere and create anything. Yet, he cared for humankind tenderly and taught the first men what they needed to know.

Edward W. Nelson (1855–1934) was a born explorer and ready for adventure. He was serving as a young naturalist and weather observer for the U.S. Army Signal Corps in Alaska when he was tapped in the late 1800s to gather information and purchase artifacts for the burgeoning Smithsonian Institution in Washington, D.C. He was living in an area he already loved for its harsh beauty and natural wonders. About his first arrival there, Nelson wrote:

> Sunday, June 17, 1877
>
> Soon the horizon to the N-E [Northeast] lighted up and changed from gray to purple then a flush of crimson touched the edges of the clouds and changed the water from a muddy green to molten copper, and as the sun approached the horizon the colors changed and shifted until the whole sky in the N.E. was one beautiful mass of colors in the form of an aurora in which the outer portion was faint crimson which became intensified toward the center while outlying masses of clouds took on changing shades of gold, which changed to yellow and then to copper and back to gray as the sun came above the waves and changed them from copper to gold. The scene was rendered still more impressive by the masses of ice which showed shades of green and blue their tops and borders silvered as they were touched by the rays of the sun—while their fantastic shapes took new form under the influence of the golden alchemist which seemed to delight in dispensing beautiful colors with a lavish hand.[8]

Nelson found himself stationed at what was the crossroads of intercontinental trade between Siberian and Alaskan Indigenous peoples. He made epic journeys along both coasts, risking his life several times. From 1877 to 1881, he met with and lived among the "Eskimos," recording their way of life, describing their ceremonies, collecting their tools and utensils, as well as making natural history observations and collections. In addition to the many myths he collected, Nelson sent back to the Smithsonian over 10,000 purchased objects illustrating all aspects of Eskimo technology, art, and culture.

## THE LITTLE CROW

East Africa. Collected and retold by Eleanor B. Heady.

All day long the black and white crows had circled over the village, cawing, cawing. Karioki, Wakai, Gachui, and some of the other boys threw clods at them to frighten them away. The girls scolded, saying, "Crows are good luck. Don't frighten them away."

When evening came and the children gathered for their story, the crows flew toward the river to roost in the treetops.

Wakai sat close to Mama Semamingi. "The girls say crows are good luck, Mama. Is that true?"

"It may be true. There is a very old story about a crow and this crow was good luck."

"Tell us, Mama Semamingi," chorused the children.

"Mzuri, good. Here it is."

Once long, long ago in the time of magic there was a good woman who had no children. She was very unhappy. "If I had a son," she said, "I should no longer be lonely. How I wish I had a child."

From the tree above her came a small voice, "I should like to be your son for I am lonely, too."

The woman looked into the tree, but could see no one except a little crow who sat on the lowest branch. "Who spoke to me?" she asked.

"It was I, Akakona," replied the crow. "I have nowhere to go. Please, may I be your son?"

"But all the people will laugh at me if I take a crow for a son," said the woman.

"You'll not be sorry. You'll find that I am a very unusual crow," declared Akakona. "Please give me a home."

So the woman took pity on the little crow, taking him into her hut and raising him as her own son. He learned to talk like the village boys. He was a great comfort to his mother.

Finally when Akakona was grown up, he saw a beautiful maiden at a nearby village. He asked her name and she replied, "I am called Kakumba, daughter of Rugeya."

Then Akakona, the little crow, sat on a tree and sang to her,

*Kakumba with the velvet eyes,*
*I'm only a crow, but I am wise.*
*Oh come with me and be my wife,*
*And you'll be happy all your life.*

Kakumba replied, "You are a handsome crow, but you are not a boy. I cannot marry a crow."

Then the little crow flew to the hut of Kakumba's father, Rugeya, and called out, "Hodi, Rugeya, it is I, Akakona. I wish to marry your beautiful daughter, Kakumba."

"But you are only a crow," said Rugeya. "Kakumba is very lovely. She must marry a handsome young man."

"I can offer a fine gift for your daughter. My mother will give me many cows. I shall bring them to you if I may marry Kakumba."

"But she would still be marrying a crow," protested Rugeya.

"You must admit I am not a usual crow," insisted Akakona. "In all ways but appearance I am a man. If you let me marry your daughter, I promise you shall not be disappointed."

"And what of Kakumba?" asked the old man. "I shall not force the daughter I love so well to marry a crow."

"Honored sir," said the crow with a bow. "Allow me to see your daughter alone for a few minutes. I'm sure I can persuade her."

"Very well. You may try, Akakona."

So the beautiful Kakumba was called to her father's hut. "Speak with the crow alone, child," he said. "I'll wait for you outside." And he left the hut.

Then Akakona, the little crow, sang,

*Kakumba with the velvet eyes,*
*I'm only a crow, but I am wise,*
*Oh come with me and be my wife,*
*And you'll be happy all your life.*

The girl replied, "You are still a crow. I cannot marry a crow."

Then Akakona hopped three times around the hut singing as he went,

*Perhaps you think that I'm a crow,*
*I look like one of course, you know,*
*But you'll discover to your joy,*
*I really am a handsome boy.*

As Kakumba watched the crow circling her father's hut she was amazed to see him growing taller as he sang. His feathers disappeared. After the third time around he finished his song and stood before her, not Akakona, the little crow, but Akakona, the handsomest boy in all the land!

"Now will you marry me?" he asked.

"Indeed I will!" gasped Kakumba. "But you must promise to remain a boy."

"I promise. The evil spell that made me look like a crow was broken by my love for you."

Akakona and Kakumba went out to her father. Rugeya prepared a great wedding feast.

Akakona went to his own village to show himself to his mother. She cried with joy that he had become a young man. Then together they drove a great herd of cattle as a gift for Kakumba's father.

There was great happiness with feasting and singing because all the people were happy for the handsome Akakona and his beautiful bride, Kakumba. Then the young people sang together,

*Akakona was a crow,*
*Now he is a boy.*
*The spell is broken. Now we know*
*Our lives shall be all joy.*

Magic spells may not keep us "spellbound" in this day and age, but not so long ago they were taken very seriously. The ancient Egyptian Book of the Dead contains 189 spells, several of which (spells 76–88) are identified as "transformation spells" to give the deceased the power to take any number of different forms in order to navigate the dangers of the Underworld. In Greece and Rome, you could use a binding spell to attract a loved one. Celtic druids often cast invisibility spells, called *feth fiada*; the incantation itself, *fith-fath*, is still remembered by some today in the Scottish Highland glens. The Old Norse word *galdr* denotes incantation spells written as runes. A sixteenth-century Hebrew manuscript, *Ets ha-Da'at*, holds 125 magic spells for all sorts of purposes: curses, healing, love, improving memory, shortening the road you are on, banishing an evil spirit. And of course, the modern story of Sleeping Beauty revolves around the fate of a princess cursed to sleep one hundred years by a spell.

**Akakona and Kakumba. 1968. Tom Feelings. Illustration for *When Stones were Soft: East African Fireside Tales.* A tale told around the world. The "handsome prince" in the form of a crow tests a young girl's love (The Tom Feelings Collection, LLC).**

The trope of a spell that traps a human in an animal form can also be found around the world. The most endearing part of this East African story is at the beginning when the old woman takes the entrapped crow to raise as her son. It seems she does not care so very much if it is bird or boy; she needs the crow to keep her company and the crow needs her. Of course, love frees the bird and everyone lives happily ever after.

The tale begins with black-and-white crows, who are thought to bring good luck, circling the village. Eleven birds of the Corvidae family call East Africa home. Several of them are indeed black and white. So it is hard to tell exactly which species is being referred to. The pied crow and the white-necked raven can easily be misidentified. Eleanor Butler Heady (?—1979) spent time in the area as an ordinary traveler who had a keen interest in local stories. She wrote in the preface to her collection, *When the Stones Were Soft: East African Fireside Tales*: "During the time I lived in East Africa, I asked everyone I met for stories." She noticed that many variations of the same story popped up in different locations: "the stories of the farming Kikuyus were heard by the campfires of the Luo fishermen on the shores of Lake Victoria or a wandering Masai told his son a Kipsigi animal story as they herded cattle on the plains."[9] In a later book, she concluded: "It is quite evident that these stories have traveled from tribe to tribe long before there was a written language."[10] Heady added that the storytellers she listened to were gifted orators, acting out their tales with dramatic gestures, snippets of song, and the startling sounds of animals. The crow confidently dipped in and out of these stories.

## The Crow

Poland. From Andrew Lang's *The Yellow Fairy Book*.

Once upon a time there were three Princesses who were all three young and beautiful; but the youngest, although she was not fairer than the other two, was the most loveable of them all. About half a mile from the palace in which they lived there stood a castle, which was uninhabited and almost a ruin, but the garden which surrounded it was a mass of blooming flowers, and in this garden the youngest Princess used often to walk.

One day when she was pacing to and fro under the lime trees, a black crow hopped out of a rose-bush in front of her. The poor beast was all torn and bleeding, and the kind little Princess was quite unhappy about it. When the crow saw this it turned to her and said:

"I am not really a black crow, but an enchanted Prince, who has been doomed to spend his youth in misery. If you only liked, Princess, you could save me. But you would have to say goodbye to all your own people and come and be my constant companion in this ruined castle. There is one habitable room in it, in which there is a golden bed; there you will have to live all by yourself, and don't forget that whatever you may see or hear in the night you must not scream out, for if you give as much as a single cry my sufferings will be doubled."

The good-natured Princess at once left her home and her family and hurried to the ruined castle, and took possession of the room with the golden bed. When night approached she lay down, but though she shut her eyes tight sleep would not come.

At midnight she heard to her great horror some one coming along the passage, and in a minute her door was flung wide open and a troop of strange beings entered the room. They at once proceeded to light a fire in the huge fireplace; then they placed a great cauldron of boiling water on it. When they had done this, they approached the bed on which the trembling girl lay, and, screaming and yelling all the time, they dragged her towards the cauldron. She nearly died with fright, but she never uttered a sound. Then of a sudden the cock crew, and all the evil spirits vanished.

At the same moment the crow appeared and hopped all round the room with joy. It thanked the Princess most heartily for her goodness, and said that its sufferings had already been greatly lessened.

Now one of the Princess's elder sisters, who was very inquisitive, had found out about everything, and went to pay her youngest sister a visit in the ruined castle. She implored her so urgently to let her spend the night with her in the golden bed, that at last the good-natured little Princess consented. But at midnight, when the odd folk appeared, the elder sister screamed with terror, and from this time on the youngest Princess insisted always on keeping watch alone.

So she lived in solitude all the daytime, and at night she would have been frightened, had she not been so brave; but every day the crow came and thanked her for her endurance, and assured her that his sufferings were far less than they had been.

And so two years passed away, when one day the crow came to the Princess and said: "In another year I shall be freed from the spell I am under at present, because then the seven years will be over. But before I can resume my natural form, and take possession of the belongings of my forefathers, you must go out into the world and take service as a maidservant."

The young Princess consented at once, and for a whole year she served as a maid; but in spite of her youth and beauty she was very badly treated, and suffered many things. One evening, when she was spinning flax, and had worked her little white hands weary, she heard a rustling beside her and a cry of joy. Then she saw a handsome youth standing beside her; who knelt down at her feet and kissed the little weary white hands.

"I am the Prince," he said, "who you in your goodness, when I was wandering about in the shape of a black crow, freed from the most awful torments. Come now to my castle with me, and let us live there happily together."

So they went to the castle where they had both endured so much. But when they reached it, it was difficult to believe that it was the same, for it had all been rebuilt and done up again. And there they lived for a hundred years, a hundred years of joy and happiness.

Andrew Lang's rendition of a Polish fairy tale is remarkably similar to the previous East African story, but it reads as a classic fairy tale. It begins with "Once upon a time," has a prince and princess, involves a call to courage, and ends with a moral resolution (the right behavior in the face of harsh challenges will bring great rewards). The handsome prince trapped in the body of an ugly crow is the centerpiece around which the action takes place.

Fairy tales gained enormous popularity in Europe after the Renaissance, particularly after the early nineteenth century publication of works by the Brothers Grimm. The brothers presented romanticized tales full of magic and morality. There were fantastical creatures and fanciful settings. The real world was described as if it were another world. So it would not have been a surprise when Andrew Lang introduced his *Yellow Book* of stories with a discussion about fairies:

> As to whether there are really any fairies or not, that is a difficult question. Professor Huxley thinks there are none. The Editor never saw any himself, but he knows several people who have seen them—in the Highlands—and heard their music. If ever you are in Nether Lochaber, go to the Fairy Hill, and you may hear the music yourself, as grown-up people have done, but you must go on a fine day. Again, if there are really no fairies, why do people believe in them, all over the world? The ancient Greeks believed, so did the old Egyptians, and the Hindoos, and the Red Indians, and is it likely, if there are no fairies, that so many different peoples would have seen and heard them?[11]

Icelanders today would certainly agree with Mr. Lang. In Iceland, fairies are ever-present and very active.

While Eleanor Heady listened to stories in East African villages, Andrew Lang and his wife Leonora collected them from secondary sources already in publication. This practice resulted in a "telephone chain" situation. For our piece, the Langs translated and cleaned up a story they had gotten which was an 1837 Slavic translation (called *Die Krähe*, "The Crow") by Hermann Kletke of a supposed Polish story. The reputed Polish original was called *Zaklety w wrone*, or "The Enchanted Crow." This version was popularized outside of Poland through Kletke's translation but was actually unknown in Poland itself. As with oral tales, the printed story changed in each iteration to suit the needs and expectations of the audiences as well as the storytellers themselves.

Scotsman Andrew Lang (1844–1912) and (more importantly) his wife Leonora Blanche Alleyne (1851–1933) are credited with having assembled over 430 tales from around the world to present to an audience who had never before heard them. Although Andrew Lang is on the cover page as editor, his wife and her female assistants were most responsible for the laborious effort of the translations and the weight of creating an engaging retelling. These were bowdlerized versions of the actual stories and intended specifically for children. The series known as *The Langs' Fairy Books* has been at countless bedsides, waiting to be opened time and time again, and read aloud to children of all ages. Each telling releases its own kind of fairy dust onto the listener.

## The Story as Told by Cornix

Ovid. Italy.

[Cornix, the Crow, explains to the Raven ...]

The famous Coroneus was my father, in the land of Phocis (it is said to be well known) and I was a royal virgin and wealthy princes courted me (so do not disparage me).

But my beauty hurt me.

Once when I was walking slowly as I used to do along the crest of the sands by the shore, the sea-god saw me and grew hot.

When his flattering words and entreaties proved a waste of time, he tried force, and chased after me. I ran, leaving the solid shore behind, tiring myself out uselessly in the soft sand.

Then I called out to gods and men.

No mortal heard my voice, but the virgin goddess felt pity for a virgin and she helped me.

I was stretching out my arms to the sky: those arms began to darken with soft plumage. I tried to lift my cloak from my shoulders but it had turned to feathers with roots deep in my skin.

I tried to beat my naked breast with my hands but found I had neither hands nor naked breast.

I ran, and now the sand did not clog my feet as before but I lifted from the ground, and soon sailed high into the air.

So I became an innocent servant of Minerva.

But what use was that to me if Nyctimene [daughter of the king of Lesbos], who was turned into an Owl for her dreadful sins, has usurped my place of honour? [The Crow was demoted for gossiping.]

The crow is here our narrator as she describes the events that led to her transformation into a bird. She had once been Cornix, virgin daughter of King Phocis, and was walking on the beach when the sea-god Neptune attempted to rape her. As she fled his clutches, the girl cried out to the goddess Minerva who rescued her by turning her into a crow. It is unclear whether this was ultimately a good thing, but the moment of transformation certainly saved her.

This theme of transfiguration is central to the whole of Ovid's monumental work *Metamorphoses* (written in 8 CE) in which ordinary human beings, as well as gods, undergo continuous change. Humans are turned into flowers, trees, animals, birds, insects, even phenomena like a constellation and an echo. The gods shapeshift into less permanent forms. Through his work, Ovid's intention was to comment on the emotional, psychological, social, and moral experiences that lead to a metamorphosis—"the feelings of that human spirit inside the changing and changed body."[12] His preoccupation was on the instability of the human condition, particularly in relation to the many kinds of "love." We visited an earlier episode from Ovid in Chapter 5 (concerning a prattling raven) where Apollo's jealous love leads him to murder his loved one. In Cornix's narrative we witness love turn into a rape scene and are relieved to see her escape.

The writer Ovid (43 BCE–17/18 CE) was supposed to become a lawyer in Rome. Instead, when still a young man, he followed all the great poets of his day, such as Virgil and Horace, and honed a special skill for writing elegies about love. He brought his prodigious skills to the creation of a sprawling tour de force that has been a source of inspiration for artists of all kinds. Since his death, hundreds of

***Neptune pursuing Coronis; Minerva interposes herself and turns Coronis into a crow.*** **1724. Francisco Vieira de Mattos. Portugal. Etching, 11.44 × 8.56 inches. A dramatic moment of rescue is achieved when Minerva (seen here with battle helmet, lance, and shield) intervenes to transform the beautiful Coronis into a crow, allowing her to barely escape the clutches of Neptune (The Elisha Whittelsey Collection, The Elisha Whittelsey Fund, 2019. The Metropolitan Museum of Art).**

painters (Picasso), sculptors (Bernini), composers (Strauss), playwrights (Shakespeare), and poets (Dante) have tapped into Ovid's renditions of the more than 250 myths that once circulated around the Roman Empire. These artists discovered the value of Ovid's wit and humor as well as his insights and artistry.

## The Man Who Was Changed Into a Crow

China. Told by P'u Sung-ling; translated by Herbert A. Giles.

Mr. Yü Jung was a Hu-nan man. The person who told me his story did not recollect from what department or district he came. His family was very poor; and once, when returning home after failure at the examination, he ran quite out of funds. Being ashamed to beg, and feeling uncomfortably hungry, he turned to rest awhile in the Wu Wang (guardian angel of crows) temple, where he poured out all his sorrows at the feet of the God.

His prayers over, he was about to lie down in the outer porch, when suddenly a man took him and led him into the presence of Wu Wang; and then, falling on his knees, said, "Your Majesty, there is a vacancy among the black-robes; the appointment might be bestowed on this man." The King assented, and Yü received a suit of black clothes; and when he had put these on he was changed into a crow, and flew away.

Outside he saw a number of fellow-crows collected together, and immediately joined them, settling with them on the masts of the boats, and imitating them in catching and eating the meat or cakes which the passengers and boatmen on board threw up to them in the air. In a little while he was no longer hungry, and, soaring aloft, alighted on the top of a tree quite satisfied with his change of condition.

Two or three days passed, and the King, now pitying his solitary state, provided him with a very elegant mate, whose name was Chu-ching, and who took every opportunity of warning him when he exposed himself too much in search of food. However, he did not pay much attention to this, and one day a soldier shot him in the breast with a crossbow; but luckily Chu-ch'ing got away with him in her beak, and he was not captured. This enraged the other crows very much, and with their wings they flapped the water into such big waves that all the boats were upset.

Chu-ch'ing now procured food and fed her husband; but his wound was a severe one, and by the end of the day he was dead—at which moment he waked, as it were, from a dream, and found himself lying in the temple. The people of the place had found Mr. Yü to all appearance dead; and not knowing how he had come by his death, and finding that his body was not quite cold, had set someone to watch him.

> They now learnt what had happened to him, and making up a purse between them, sent him away home. Three years afterward he was passing by the same spot, and went in to worship at the temple; also preparing a quantity of food, and inviting the crows to come down and eat it. He then prayed, saying, "If Chu-ch'ing is among you, let her remain."
>
> When the crows had eaten the food they all flew away; and by-and-by Yü returned, having succeeded in obtaining his master's degree. Again he visited Wu Wang's temple, and sacrificed a calf as a feast for the crows; and again he prayed as on the previous occasion. That night he slept on the lake, and, just as the candles were lighted and he had sat down, suddenly there was a noise as of birds settling, and lo! some twenty beautiful young ladies stood before him.
>
> "Have you been quite well since we parted?" asked one of them; to which Yü replied that he should like to know whom he had the honor of addressing.
>
> "Don't you remember Chu-ch'ing?" said the young lady; and then Yü was overjoyed, and inquired how she had come.
>
> "I am now," replied Chu-ch'ing, "a spirit of the Han river, and seldom go back to my old home; but in consequence of what you did on two occasions, I have come to see you once more."
>
> They then sat talking together like husband and wife reunited after long absence, and Yü proposed that she should return with him on his way south. Chu-ching, however, said she must go west again, and upon this point they could not come to any agreement. Next morning, when Yü waked up, he found himself in a lofty room with two large candles burning brightly, and no longer in his own boat. In utter amazement he arose and asked where he was.
>
> "At Han-yang," replied Chu-ch'ing; "my home is your home"....

The character Yü carries on with his life in a long, winding narrative about wives, children, love, loyalty, and home. When Yü feels like traveling, he takes out his black clothes (the typical clothing of Buddhist monks), puts them on, feels wings immediately grow from his ribs, and "with a flap he is gone." At first, it is an astonishment—and a blessing from the emperor/god Wu Wang—later it is a great convenience. In the chronicle, the border between crow and human is fluid. Yü appears as both, at different times; his wife Chu-ch'ing also easily changes form when needed. The feeling of the story itself is that of motion—of flying smoothly back and forth, through space, time, habitations, and bodies. Jorge Luis Borges picked this story as one of his favorites of all time.[13]

"The Man Who Changed Into a Crow" appears to have been a bit biographical. *Liao-Chai-Chih-I*, or *Strange Stories from a Chinese Studio*, was written in the seventeenth century by Pu Songling (1640–1715, identified as P'u Sung-ling by Mr. Giles). Mr. Pu came from a recently impoverished middle-class family in Shandong Province, on the northeastern coast of China. Like Mr. Yü, he studied diligently to pass the imperial examinations but never succeeded past the county level and so could not secure an appointment as an imperial bureaucrat. Pu turned to collecting folk tales and telling fantastical stories, producing 491 tales in sixteen volumes whose pages were filled with ghosts, flower-spirits, fox-fairies, and other assorted beings. The plots of these "marvel tales" are fundamentally romantic and philosophical in nature. They explore the ramifications of what the term 情,*qíng*, connotes: warmth, love, passion, kindness, leniency, favor.

Pu's manuscripts were copied and recopied as they were passed around to friends and family. He was too poor to pay for the expense of block-printing. It was not until after the author's death that the volumes were committed to print by his grandson. Herbert Giles (1845–1935), a British diplomat and sinologist of the Victorian era, was eager to introduce Pu's works to an English audience. He served in China for over twenty years and completed a comprehensive Chinese-English Dictionary as well as a history of Chinese literature and art. His translation of *Liao-Chai-Chih-I* leaves out or changes the abundant salacious material that is found in Pu's stories but it nevertheless retains the work's dreamy exuberance.

## Tempo

Noreen Lawlor. USA.

I am a crow grown tired of my
wings
beating the vastness of sky
under them
I want to be of earth of the
cringing soil
and the scalding bones of men
I want to be of darkness not of
feathers
no longer fill my nest
with shiny things
rings from cracker jack boxes
it is the time
of hair and hands and blue eyes

In a poem that reads like beating wings, Noreen Lawlor inhabits the body of a crow to give us a crow's point of view. This crow wants to transform into something else. It is tired of the vastness of the sky. It is tired of its feathers and tired of shiny trinkets. The poet as crow says that it is time ... to become human.

Being human seems to be as primal as being a crow. A human is made up of hair, hands, and blue eyes (not more remarkable things like song, dance, or poetry). A human inhabits cringing soil and darkness; its bones are scalded by the sun. We are left to wonder why a bird capable of soaring would wish to be grounded on earth.

***King of Birds*. 2012–2013. Photograph from the series "Eisbergfreistadt." The artistic team Kahn + Selesnik revel in the unreal—creating installation art that is fictitious but almost possible. They disrupt our way of seeing and thinking in ways very much like mythic storytelling. A man is engulfed by crows. His head is obscured, possibly gone. He appears to be changing into something else. The image is unsettling. The man stands frozen while the birds upon him are busy with activity. Nature may be reclaiming the human form (Courtesy of Nicholas Kahn & Richard Selesnick).**

The poem, which begins in the sky then descends to earth, hints at a metaphorical reading. Perhaps the crow is the poet's ambitions. Perhaps venturing across the world looking for sparkling enticements is exhausting and ultimately fruitless. The bird may want to leave behind crow-like (childish) things and take up the work of becoming a human adult. Hair, hands, and blue eyes seem to be more important than any bright objects.

Noreen Lawlor lives in Joshua Tree, California.

11

# In Ceremony and Performance

The crow has given me the signal,
The crow has given me the signal,
When the crow makes me dance,
When the crow makes me dance,
He tells me (when) to stop,
He tells me (when) to stop.

To conclude an Arapaho Ghost Dance in the nineteenth century, the crow messenger would descend to earth and confirm that those in the Beyond had heard the prayers of the living, had listened to their songs, and had participated in their trance journeys. Thus satisfied, the dancers would break their circle to shake their blankets and shawls in the air and drive away all lingering evil influences. They then went to wash themselves of the same.

Even today, the vehicle of an orchestrated ceremony impels the imperceptible forces around us to appear in a form we can understand. Our visions come to three-dimensional life amid the rhythmic shuffle of feet, the steady beat of drums, the melody of voice, and lingering lines of movement. An illusory world becomes real; the real becomes visible. We are then able to enjoin the forces—to cure, to educate, and to align ourselves with the broader universe. Ceremony and performance have been a part of the human experience since our beginnings. They are powerful confirmations of what we know and catalysts for new perceptions. The ritual intricacies of a Catholic Mass, for instance, are deeply familiar to all Catholics but nevertheless lead to epiphanies and revelations on all matter of religious understanding. At Passover, Jewish families place an extra glass of wine on their seder table and open the door to their homes for the prophet Elijah to enter and drink. They feel his spirit, and all he represents, alive among them. Rituals make meaning. Ceremonial theater serves community and strengthens bonds as it facilitates the expression of our deepest values. The crow and raven are called upon to participate in many of these rituals.

## *Field Notes: Courtship*

In nature, the crow and raven do not express any "values" but they too are performative animals. Their rituals of courtship are particularly well choreographed

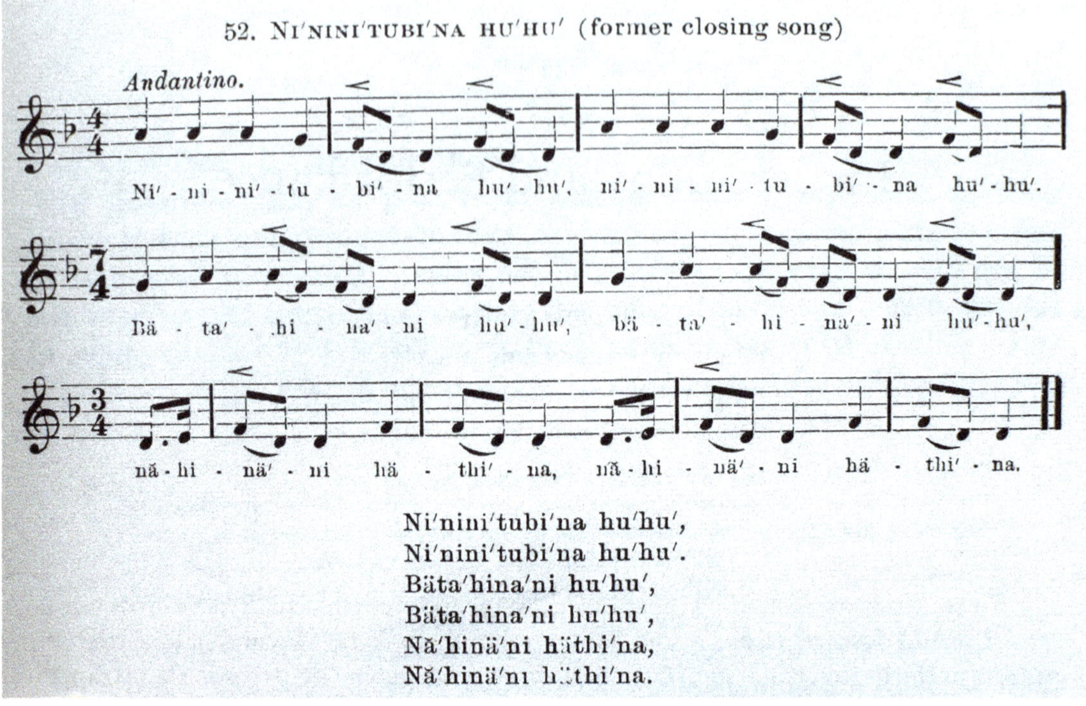


**Arapaho Ghost Dance closing song (see translation, on previous page). Anthropologist James Mooney published musical transcriptions for more than a dozen sacred songs involved in the Ghost Dance ceremony. From "The Ghost-dance religion and the Sioux outbreak of 1890," James Mooney, 1897. *Fourteenth annual report of the Bureau of Ethnology, 1892–1893*: 641–1110. Smithsonian Institution Bureau of American Ethnology. Washington, D.C.: Government Printing Office.**

gestures of intention which in turn create lasting bonds. Corvids are late bloomers, but early season breeders. They wait until they are between two and four years of age before breeding. The beginning of March is when the action gets going.

Unlike many birds, crows and ravens engage in a long, drawn-out courtship pattern before pairing up. Ravens spend a lot of time checking each other out and then they might pair up as friends first. Bernd Heinrich writes that they can begin to court when only about six months old even though mating does not take place until they are at least three years old. And courting can be a year-round activity for unmated dominant birds (those are the ones with the best chance at success in partnering).[1] Young birds will try to establish their dominance through acrobatic flight displays, by collecting trinkets to show off (such as shiny objects or feathers), and by leading others to good food sources. Dr. Heinrich noticed that, among his raven-subjects, dominance was broadcast through a display of grand gestures—puffed up strutting with head held high and bill angled up, flashing white nictitating eye membranes, elevated "ears," a greatly broadened neck, and extended throat hackles. Usually challengers backed down, but if not, a fight might ensue.[2]

Once the males get serious about the opposite sex, their pre-mating performances will include any of the following to impress a female: For ravens—dramatic

**Raven courtship in Spain. Corvids engage in many courtship displays that will cement a lifelong relationship. Touching beaks, preening, and many kinds of vocalizations are just a few ways they express an interest in one another (George Reszeter/Alamy Stock Photo).**

sky dives and rolls (sometimes with a food delivery added),[3] vocalizations such as gentle "coo"s or gurgles, or a rattle sound accompanied by percussive bill-snapping. Ravens will solicit and offer preening opportunities to just about anyone at first but then gradually become more specific. For crows—"singing" from a tree branch or on the wing, executing jittery routines of head-bowing while pitching side to side, vigorous bobbing on a tree branch, strutting along the ground with spread wings and tail, or even flattening the whole body to the ground and fanning out the tail. Of course, fluffing out one's body feathers to the max is very attractive to all corvids.

As soon as the female has assessed the qualities of a potential mate (upon whom she will depend completely during nesting) and given her approval, a period of mutual attention follows, with bonded pairs spending time flying together and preening each other. They will cement their relationship with just one individual through very intimate preening—running a beak through the neck and chest feathers of the mate. This holds true for crows too. A male will be encouraged to rub his mouth and nose against a partner in a kind of nuzzle. They might then engage in a "dance" where male and female fly up to each other and touch beaks. They will perform an aerial loop dance together, mimic each other's movements in mid-air, and sometimes lock talons. The display can last for several minutes before the couple finally lands.[4]

The birds will build a nest together and finally mate. They stay a couple for life, and during this time they continue their demonstrations of affection—gentle bill grabbing, grooming, and soft utterances. In the woods beyond my backyard, I once saw a pair of crows standing on a fallen tree trunk. They were facing one another and bobbing up and down. On the downstroke, each "sharpened" its bill on the trunk—first a left swipe of the bill, then a right swipe. Was this lovely activity a display of their affection for one another?

People who are tuned into the ways of nature have surely paid attention to the intricacies of corvid behavior. In performances around the world, we can see avian movements appropriated to bridge the gap between human and bird. When a man behaves like a bird, he becomes something greater than himself. The aura of mythic power that the bird represents is channeled and conferred onto sacred ritual observances. We take a look at just a few of the many that the crow and raven have inspired.

## Raven Dance

Yup'ik, Bering Strait. Described by Lieutenant Lavrentiy A. Zagoskin; translated by Penelope Rainey.

The people sit on all three tiers of benches and on the floor except for the front side, which is left free for the performers. The men occupy the benches; some are without their parkas, some altogether naked. It is hot and stifling. Two oil lamps on the "proscenium," that is, in the corners of the front side of the fire pit, and four additional ones in different parts of the kazhim [large communal house] throw a dim light on the motley crowd of spectators. There are grass mats hanging from the front edge of the lowest benches and they separate the actors' dressing-room. Four shamans are sitting on this bench holding drums two feet in diameter in their teeth. Two old men in tattered parkas with smeared faces appear on the stage from time to time, tease each other and make fun of the spectators, saying that the latter in vain have come together hoping to see the new dance which they themselves, the old men, stole from the man who made it up. This is in place of an overture.

But now the skylight opens, and quickly, in a flash, a dancer slides down a strap and with a quick leap is on the stage; two pairs of women take their places beside him. He is wearing a mask representing a fantastic raven's head, and there he goes jumping about on the stage, calling like a raven; the drums sound their rhythmic beat, the singers strike up a song. The dancer at one time represents a raven perching and hopping like a bird; at another time he represents the familiar actions of a man who is unsuccessful in everything. The content of the dance is explained in the words of the song and may briefly be described as follows. A shaman is living in his trail camp. He is hungry, and he notices that wherever

> he goes a raven goes with him and gets in his way. If the game he is pursuing is a deer, the raven from some place or other caws, startles the deer, and makes it impossible to creep up within a bow-shot of it. If the man sets a noose for hare or partridge, the raven tangles it or runs off with it. If he sets a fish-trap for imagnat [?], there too the raven finds a way to do him harm. "Who are you?" cries the shaman at last. The spirit in the form of the raven smiles and answers: "Your evil fate." Thus in this dance are combined the mimicking of a deer hunt, the catching of hares or partridges, and of fish, and a conversation between the shaman and the raven.

Raven appears in dances all along the Northwest coast of the Americas. His presence is invoked by shamans and clan chiefs. He is also addressed by ordinary hunters and fishermen. His participation in the human world is made manifest in elaborate scenes of song, drumming, and masked movement where Raven is revealed for who he is—a trickster, a creator, a lover, a disrupter. He is in many ways the force of creative chaos—his powers blowing in every direction, toward good luck and bad luck, and sometimes both at the same time. When asked "Who are you?" Raven answers "Your evil fate." We don't know what to make of it.

Among the Yup'ik of the subarctic coastal tundra, the kazhim (or *qasigiq*) was where the most important rituals took place. This large men's house hosted ceremonies that were both deeply sacred and very entertaining. Partially sub-terranean and constructed to include wide benches and an open performance area with a large fire pit in the middle, the kazhim was perfect for theatrical staging effects that engaged all the senses. Shamans composed much of the music and dances, and they oversaw the making of masks designed specifically for each event. The masks were especially important because they represented shamanic visions and empowered the shamans to take on a spirit's character and speak through its voice. The dance was the communication in mime of how the spirit appeared to the shaman, what it did, and what it said. At the end of the event, most masks were burned or buried to rid them of their potentially dangerous power and to send them back to the sacred world.

When Lavrentiy Zagoskin (1808–1890) first encountered the Yup'ik in the early 1840s very little was known by outsiders about the native peoples of the area. Zagoskin, a Russian naval officer, had been sent by the Russian government and the Russian American trading company to uncover fur-trading opportunities. He was to map the geography, assess the resource wealth, and make contact with the inhabitants so that the Russians could intercept established aboriginal trade routes and maximize profits in increased trade. He covered an enormous territory by foot, snowshoe, and skin boat and produced the first ethnographic reports on the peoples living there. Fortunately, he was a diligent witness and careful recorder who knew the limits of his work. He deliberately reminded readers that language constraints prevented him from understanding much of the social and religious customs that he observed. He did not pry into areas that were culturally sensitive, preferring instead

to simply record what he saw. Zagoskin kept voluminous notes about his eleven trips over two years of expansive travel.

The United States purchased Alaska from the Russian Empire in 1867. American Edward Nelson (Chapters 3 and 10) followed in Zagoskin's footsteps in the late 1890s. While Zagoskin was after Yup'ik

***Above and left:*** **Raven dance. Left: Carving, Dancer and Drummer. ca. 1970. Leo Kunnuk Jr. Ugiuvak (King Island, Inupiaq territory). Carved walrus ivory mounted on baleen, 24 × 55.25 × 27.75 inches (permission courtesy of the artist's family; Anchorage Museum, 1996.7.51). Right: Contemporary Raven performance by the Chilkat Dancers of Haines, Alaska (Patrick J. Endres/Alaskaphotographics.com). In the North, winter is the time for dance and performance. Among the Native peoples of Alaska, the Raven Dance has been passed down from generation to generation. The stories may change but the importance of the raven does not.**

furs, Nelson was after their cultural objects. The Yup'ik people survived the Western onslaught, nevertheless. Today, their stories and dances remain cherished, and their spiritual and cultural ways are intact. The Raven dance is taught to youth and performed with a special pride of heritage. The communities of the ancestors, the animal spirits, and the living are bound together through these demonstrations of belonging.

## Calling the Medicine Poison

Pit River (Achomawi), California. Told by Jaime de Angulo.

That evening, we all gathered at sundown. Jack Steel, an Indian from Hantiyu who usually acted as Blind Hall's "interpreter," had arrived. He [Blind Hall] went out a little way into the sagebrush and called the poisons. "Raven, you, my poison, COME! (*qaq, mi,' ittu damaaqome, tunnoo*).... Bullsnake, my poison, come.... Crablouse, my poison, cooome.... You all, my poisons, COOOME!!" It was kind of weird, this man out in the sagebrush calling and calling for the poisons, just like a farmer calling his cows home.

We all gathered around the fire; some were sitting on the ground, some were lying on their side. Blind Hall began singing one of his medicine-songs. Two or three who knew that song well joined him. Others hummed for a while before catching on. Robert Spring said to me: "Come on, sing. Don't be afraid. Everybody must help." At that time I had not yet learned to sing Indian fashion. The melody puzzled me. But I joined in, bashfully at first, then when I realized that nobody was paying any attention to me, with gusto.

Blind Hall had soon stopped singing, himself. He had dropped into a sort of brown study, or as if he were listening to something inside his belly. Suddenly he clapped his hands; the singing stopped abruptly. In the silence he shouted something which the "interpreter," Jack Steel, repeated. And before Jack Steel was through, Blind Hall was shouting again, which the interpreter also repeated, and so on, five or six times. It was not an exchange between Blind Hall and Jack Steel. Jack Steel was simply repeating word for word what Blind Hall was shouting. It was an exchange between Hall and his poison, Raven. First, Hall would shout a query which the interpreter repeated; then Hall would listen to what Raven (hovering unseen above our heads) was answering—and he would repeat that answer of Raven which he, Hall had heard in his mind—and the interpreter would repeat the repetition. Then Hall emitted a sort of grunted "Aaah ...," and relapsed into a brown study. Everybody else, Jack Steel included, relaxed. Some lit cigarettes; others gossiped. A woman said to me: "You did pretty good; you help; that's good!" Robert Spring said: "Sure, everybody must help. Sometimes the poisons are far away. They don't hear. Everybody must sing together to wake them up."

Jamie de Angulo (1887–1950) begins what would become a cult best-seller, *Indians in Overalls*, in this fashion: "A STREET in a little town on the high desert plateau of northeastern California. Clear air, blue sky, smell of sagebrush, smell of burning juniper wood. I was looking for Jack Folsom, an Indian of the Pit River tribe." de Angelo was a medical doctor-turned-self-taught ethnologist who wandered among the Achomawi (meaning "river people" but called Pit River people by Westerners) collecting linguistic information and stories. de Angulo learned about an old fellow named Blind Hall (or Tahteumi) who had a reputation for being a powerful medicine man. He found Blind Hall (who was indeed blind) to discover that the man was himself sick. Blind Hall told him: "I am pretty sick now, dropped my shadow on the road, can't live without my shadow, maybe I die, I dunno.... I doctor myself tonight. You stay. You help sing tonight.... I got several poisons.... I got Raven, he live on top mountain Wadaqtsuudzi, he know everything, watch everything."[5] de Angulo stayed for the healing event, which lasted a couple of hours. The next morning Blind Hall felt much better. His Raven must have helped.

The Achomawi did not have a formalized religious practice, but they did have stories and songs, dreams and "poisons" (powers). Every entity or object in the Achomawi world was full of life—a rock, a tree, a snake, the sun, or the stars could be conversed with or enlisted to be a spirit guardian. Shamans ("medicine men") acquired diverse spirit guardians and, through them, the ability to bring about healing or sickness. Spirits could make things happen—both good and bad. The singing of a spirit's power song called down the spirit. If the spirit sang its own song but went unanswered, a person would become sick. The shaman was then needed to figure out what the song was and how to sing it.

Blind Hall mentions losing his shadow, his *delamdzi*. This was a substance made of light that left a person during sleep to go over the mountains and return at dawn. de Angulo writes: "A man who has lost his shadow is said to be alive, but that is only an appearance. He is really half dead. He can last like that for a few days, even one or two weeks."[6] The singing of songs summoned the "poisons" and, along with other treatments, brought about healing.

Jaime de Angulo was as flamboyant a character as you would ever want to read about. He was a cross-dressing, philandering, self-described "freedom-loving anarchist" and alcoholic who was notorious for riding his horse naked and shouting into the night to invite the spirits to meet him at the ridge crest. One acquaintance wrote:

> His appearance, in the 1920s when I first saw him, was dramatic in the extreme. He came riding down our hill to Rainbow Lodge [in Bixby Canyon] on a black stallion, wearing black chaps, a black shirt and a black sombrero, along with a huge turquoise studded Indian silver conche belt from New Mexico. His long black hair flowing in the wind, his blue eyes flashing, he was beautiful rather than handsome and was given to passionate gestures, speaking with his hands as well as his tongue. And he talked rapidly, brilliantly, usually about linguistics, the American Indians, or Freud. He tried to make love to my mother and called her bourgeois when she refused.[7]

de Angulo nevertheless published several works. *Indians in Overalls* was written about his first linguistic fieldtrip to the Achomawi in 1921 but was not published until just before he died in 1950. After his death, de Angulo became a cult figure very much in the Jack Kerouac mold.

## Angwusnasomtaqa, Crow Mother Kachina

Hopi, Arizona. Described by Edward A. Kennard; illustrated by Edwin Earle.

On the ninth day, every year, three of the most important Kachinas make their appearance early in the morning. *Angwusnasomtaqa* [Crow Mother] goes to a shrine just north of Oraibi where she is met by the Powamu Chief. From there the Kachina gives her long-drawn-out call in the direction of the village. The Chief places prayer sticks, prayer feathers, and meal in the shrine, finally "making the road" from the east to the shrine and from the shrine to the village. At several places along the trail he makes the cloud and rain symbol, and then hands the Kachina a woven tray containing fresh green corn shoots that have been grown in the kivas. Angwusnasomtaqa proceeds toward the villages periodically giving her call. As she passes by the houses women and children sprinkle her with meal and take a shoot of green corn or a sprig of spruce from the tray. With great dignity, holding herself erect and looking neither to left nor right as the people approach, the Kachina passes through the village and eventually stops on the east side of the Powamu kiva, where she chants a long prayer.

Meanwhile, in another kiva, *Eototo* [Nature/Father/Chief] and *Aholi* [Chief's Lieutenant] have been preparing. Both of these Kachinas carry a chief's stick, shaped in a distinctive way, a netted gourd, and several bunches of green corn shoots. The chief's stick is a sacred object, but it also serves to indicate the chiefly position of Eototo and Aholi in the Kachina world. They emerge and go to the kiva where Angwusnasomtaqa is waiting, Eototo leading and Aholi following. On the way Eototo makes the triple cloud symbol with meal and Aholi places the butt of his stick upon it. Then he gives his call and waves the upper end in two circles. He faces in the opposite direction and repeats the action. Then, they proceed to a place where a hole has been dug in the plaza and a prayer stick put in it. Eototo "makes the road" with meal from the north to the hole and pours water from his gourd into it from the same direction. He does the same from the other five directions (above and below are counted as well as the cardinal directions). Aholi duplicates all his actions. Then they both take up positions beside Angwusnasomtaqa. Here Eototo again makes a line with meal from the north to the side of the kiva hatchway and pours water into it from his gourd. He repeats from the other three sides in order and Aholi does the same.

> At this point the Powamu Chief, Kachina Chief, and several members of the Powamu society come up the ladder dressed in ceremonial clothing of kilt and sash, their hair down full length, and prayer feathers tied on the top of their heads. One blows smoke upon each of the three figures, another sprinkles medicine upon them with a long feather, and all give them prayer sticks and prayer feathers. First Angwusnasomtaqa is given the offerings and then disappears from the village to deposit them in the Kachina shrine. Then, Eototo and Aholi are blessed and given prayer offerings while some of the green corn plants are taken from them.

The Hopi people ("Peaceful Ones") live in a sacred place. After emerging from under the earth at the beginning of human time, they migrated to an area near *Tuwanasnsayi*, "Center of the Universe," to establish their communities upon dry, rocky mesas in northeastern Arizona. Each of their nine villages is laid out around an open plaza and each harbors a number of *kiva*s, underground ritual spaces. From the kivas emerge *katsinam*, personified ancestral spirits. Katsinam (or katchinas/kachinas) come to the villages to dance and sing, to listen to prayers and petitions, and to bring presents to children and rain to the community. The number of katsinam is so large that it is uncountable; everyone who dies becomes a katsina. These ancestral beings are believed to live in the San Francisco mountains. They visit the villages for half the year, during which a series of ceremonies, ritual dramas, and performances are held to bring fertility and rain.

*The Crow Mother katsina here early in the morning*
*Sings and sings her songs here*
*So that life will be lived according [to the Hopi way].*
*Heee hee'e'eehee,*
*[She] keeps singing and singing.*

song sung as Angwusnasomtaqa approaches the plaza[8]

Angwusnasomtaqa, "The one with the crow tied on," is a prominent katsina. She is considered to be the mother of all katsinam by many Hopi. She appears twice (played by a man). Once, at the initiation ritual of all Hopi where she oversees the whipping of ten-to-fifteen-year-old boy and girl initiates to ensure that they remember (and keep secret) the imparted lessons of their culture, and a second time at the Powamu ceremony, which we read about here. The Powamu ritual is a prayer for fertility—of women and of the earth. Dressed in bridal clothes, Angwusnasomtaqa enters the plaza singing. She brings with her corn and bean plants that have been newly grown within the womb of the kiva. She distributes nature's gifts to those who approach her. She endows life. Dozens more katsinam appear and dance and distribute gifts. The rhythmic sound of the bells and rattles on their stamping feet beckon the "cloud fathers." Feasts are held, and at around midnight the katsinam return to their kivas to dance again. When all is over, they remove their headdresses and either place them in a shrine or plant them as a prayer for rain.[9]

In December 1935, artist Edwin Earle (1904–1989) arrived at the Hopi village of Oraibi. He teamed up with anthropologist Edward Kennard (1907–1989) to put together a beautifully illustrated book about Hopi katsinam. Dr. Kennard had been hired by the Federal Writers' Project of the WPA (Works Progress Administration) to produce a series of books about Indigenous culture and archaeology. He was conscientious in his efforts and careful with terminology. He rejected a book on Nevada's native inhabitants because it referred to "bucks and squaws" instead of "men and women," writing: "Whatever may be common usage among ranchers and others, in the American Guide Series we do not speak of 'buck' and 'squaw' and this is a fixed policy to which there will be no exceptions." He further reminded the author of that book that "'Princess' is an inaccurate term to describe social position among North American Indians.'"[10]

*Hopi Kachinas* was originally published in 1938. It remains an important source of information about some of the performative aspects of Hopi life.

**Angwusnasomtaqa. 1971. Edwin Earle. Illustration, Plate VI for *Hopi Kachinas*. Crow wings are tied to each side of the mask and white eagle feathers, known as breath feathers, stand on top. She has a collar made of spruce branches and wears a white wedding belt. The robe is a bride's blanket whose borders are embroidered with rain symbols. Angwusnasomtaqa wears a Hopi loomed dress and white buckskin moccasins. She holds corn and bean plants that have been grown in the kiva.**

## The Raven Crown

Bhutan. November 6, 2008, Simon Denyer for Reuters.

THIMPHU (Reuters)–With medieval tradition and Buddhist spirituality, a 28-year-old with an Oxford education assumed the Raven Crown of Bhutan on Thursday, to guide the world's newest democracy as it emerges into the modern world.

As the chief abbot chanted sacred sutras to grant him wisdom, compassion and vision, Jigme Khesar Namgyel Wangchuck was crowned Bhutan's Fifth Druk Gyalpo, or Dragon King, by his own father, who imposed democracy and then abdicated two years ago.

Dressed in a red and gold gho—the knee-length gown all Bhutanese men wear—he then sat cross-legged on the ornate Golden Throne, looking solemn but allowing himself one fleeting smile, as offerings were made to the new king and the gods.

The red and black silk crown, embroidered with images of white skulls and topped with a blue raven's head, represents Bhutan's supreme warrior deity and a monarchy that united this country 100 years ago and remains enormously popular.

The crowning of a new king is world news in any era. For the people of Bhutan, it was a momentous occasion in 2008. A young, worldly king bowed his head to receive the Raven Crown which would empower him to act as a personification of the supreme warrior and protector deity, Mahakala. The king of Bhutan would become a *Dharma* king—guardian of his people and of Dharma (divine law).

It is told that in 1616 CE Ngawang Namgyal, a 23-year-old Buddhist priest, was visited by the powerful god Mahakala in the form of a raven. Mahakala guided Ngawang to Bhutan to teach Buddhism and establish a theocratic Bhutanese state. Since that time, the raven has been a national symbol and the raven-faced Mahakala, the supreme defender of the land. The Raven Crown, designed by Lam Jangchub Tsundru, was first worn around 1885. Every king has since been enthroned with it. It takes the form of a traditional battle helmet that was thought to have magical, tantric powers for the subjugation of evil forces. The design and purpose of the battle helmet evolved over time, becoming the model for the royal Raven Crown which now symbolizes the sacred nature of Bhutanese kingship.

The confrontational deity, Mahakala, takes many forms. The Raven-headed Mahakala, known sometimes as Legön Jarok Dongchen (Lord of Action), is recognized throughout central Asia as a valued and ferocious preserver of cosmic order against the threats of hatred and destruction. Painted in many scrolls and mandalas, the symbolism of his black features is potent, suggesting his role

**Left: The new king of Bhutan is enthroned with the raven crown. 2008. The Raven Crown has been worn by all Bhutanese kings since the beginning of the twentieth century. It is made of embroidered silk and wool and is believed to bestow the awesome powers of the deity Mahakala onto the wearer (courtesy of the Royal Office for Media, Bhutan). *Above:* Raven crown. On the raven's head sit the sun and the moon, symbolizing longevity, steadfastness, and enlightenment. From its neck flow threads of red (drawing by author).**

over darkness and death (after all, the raven is a swift, sharp-beaked, devourer of flesh). His presence is a reminder that the Void, the darkness, is the absolute reality. The raven-headed Mahakala pursues wicked souls and enemies of the Buddhist faith and, with his sharp weapons, scares away the demons and evil within and around us.

The Raven-headed Mahakala also features in Tibetan, Mongolian, and Bhutanese *cham* dances (illustration on following page) which are performed by colorfully dressed, masked monks at important times throughout the year. As a fierce manifestation of Avalokiteshvara, the Bodhisattva of Compassion, this Mahakala is the most popular of the Protector Deities. The main purpose of a cham dance is to purify a ritual space and destroy "obstacles" of every kind.[11] What better deity could there be than this one? His fearsomeness works to our benefit. He brings us face-to-face with death but strips away our attachments and our self-illusion (the biggest obstacle of all). The cham dance is fundamentally a danced meditation. The body movements are sweeping; performers bend at the waist and whirl their extended arms in wide arcs. They make small hops and leaps as they create intersecting loops in space. These dances are broken up by comic skits performed by clown characters and by the central event, the violent destruction of the *linga* (evil forces or obstacles in the shape of a barley-flour and butter effigy). In this way, the Raven-headed Mahakala protects us and guides us.

**Performer at Katok Dorjeden Monastery in Tibet. Many deities appear in the monastery's courtyard during the *Monlam Chenmo,* Great Prayer Festival, Tibet's most important Buddhist celebration of the year. Old costumes were confiscated and destroyed during China's Cultural Revolution and had to be reconstructed from memory and with the help of private and museum collections (Craig Lovell / Eagle Visions Photography / Alamy Stock Photo).**

## Song of Kararat (Carrion Crow Goddess)

Ainu, Japan. Translated by Donald L. Philippi.

I lived
in the Upper Heavens,
dwelling
among the gods.
However,
whenever I would hear
the sounds of feasting,
the sounds of drinking
of the gods
who had received [presents of]
human *inau*
and human wine,
I would always
be longing to have them.

I longed for them
so very much that,
when I would get lonely,
I would stand up and
would do
the dance of the glittering treasures,
the dance of the glittering metals
on the floor at the head of the fireplace.
Then acorns would come dropping down
from one of my hands,
and chestnuts would come dropping down
from my other hand.
Thanks to this,
I was able to amuse myself,
and this was the way I
continued to live
on and on and on
uneventfully.

Then news spread
among the gods

that I was doing this,
and only then
did the gods
become aware for the first time
of my existence.
After that,
when wine was delivered
from the humans,
I was invited for the first time,
and I was able
to attend a drinking feast.

I drank, and
o how very delicious
was the wine!
As I drank,
my heart
was very
mellowed by the wine,
mellowed by the liquor.
At that time
I did
the dance of the glittering treasures,
the dance of the glittering metals,
moving up along the floor
and down along the floor.
As I danced,
acorns came falling down
from one of my hands,
and chestnuts came falling down
from my other hand.
Then the gods
began to race each other
to pick up
the chestnuts
and to pick up
the acorns
on the floor.
Sounds of loud laughter,
sounds of great merriment
rose up all at once.

While this was going on,
the God Ruling
the Upper Heavens
spoke these words:
    "I did not know
    until now
    that the weighty deity
    the *kararat* goddess
    had her dwelling
    so very close by,
    near my own house.
    One of the reasons
    why I invited
    the *kararat* goddess
    was because I wished
    to apologize to her,
    but look
    how mellowed by the wine
    are her spirits!"
Thus did he speak.

The peerless feast
wore on to its conclusion.
After that,
I have remained
in my own house.
Ever since then,
whenever the gods
are worshiped
by the humans
and wine is delivered,
there is not a banquet,
not a drinking feast
from which I am
ever omitted.
I am always invited
and attend every feast,
and as I drink,
my spirits are
mellowed by the wine.
After that
I do

the dance of the glittering metals,
the dance of the glittering jewels
among the guests at the feast,
and acorns fall down
from one of my hands,
and chestnuts fall down
from my other hand.
The gods
race each other
to pick up
the chestnuts
and the acorns.
Sounds of loud laughter,
sounds of great merriment
rise up all at once,
and I take
delight in all this
as I attend
all the noble drinking feasts, the noble feasts.

This is the way
I continue to live
on and on.
Whenever wine
or inau
are delivered
from the humans,
I am given
portions of *inau*
and portions of wine,
and this enhances my glory as a deity.
This is the way
I continue to live
on and on
uneventfully.

*Kararat*'s dance of the glittering treasures must have been (and possibly might still be) a delightful sight for the assembly of Ainu gods who populate the heavenly world. We heard about the exploits of the Owl God on Earth in Chapter 5, now we hear a first-person account of the Carrion Crow goddess's activities above. The goings-on, in fact, were not much different from those of humans below. Ainu gods were very much mirrors of human beings (except they were supernatural and very powerful). They lived the same way and liked the same things. They often lead "uneventful" existences but loved to dance and sing and would cloak themselves in clouds, or appear as a plant or animal, in order to attend human events. The gods also coveted "treasures" in very human-like ways, particularly gifts of wine and *inau* (sticks of willow beautifully whittled with curled tufts). A great variety of inau was therefore made by villagers for different deities and for different rituals. The god who received many gifts was admired by the other gods. His or her prestige increased. If their "house" was filled with gifts from humans, they gave a party and celebrated with laughter and drinking. For this reason, the gods went on "business trips," much as people did. In return for presents of inau and wine, the gods gave animal flesh and fur, or blessings, or health.

This *kamuiyukar* (mythic epic) was performed for an eager audience. It was set down in writing in 1932 by Kubodera Itsuhiko from the reciter Hiraga Etenoa whose artistic talent made her "the best Ainu epic reciter whose repertory has ever been recorded in writing."[12] Donald Philippi, who lived in Japan from 1957 to 1970, worked meticulously with Dr. Kubodera to construct an English translation that conveys a sense of the original style, the unfolding of action, and the rhythm of speech. Each verse contained a burden (not included here), a refrain repeated over and over that rendered the performance even more hypnotic and enveloping.

## Crow Two: A Ceremonial Opera

Description by Pauline Oliveros, U.S.A. 1974–1975.

What happened was that I began to compose with the meditations; taking them and making situations where there were different groups doing different meditations simultaneously, making a layered composition. I wrote a piece called Crow Two.... It was quite complicated in its arrangement of meditations. It had a visual meditation: a mandala formation which was made of a large circle of people. A white-haired elderly woman was the Crow Poet and sat at the mandala center. At the four points of the compass there were two white-haired and two black-haired women—who were the four Crow Mothers. These were women chosen for their presence and the color of their hair. On the outside of the circle—at the mid-points of the compass—were four dijeridoo players. The dijeridoo is an Australian aborigine instrument. The meditation began with the drummers—there were seven drummers—doing what is called a Single Stroke Roll meditation, which is to imagine the equal alternation between hands or mallets on the instrument and to let the roll begin from the imagination. The body responds to the imagination and starts the roll rather than will it. If that instruction is carried out faithfully, the roll does indeed begin but it is involuntary and it locks onto some

> internal body rhythm. They are instructed not to change anything, but to keep it going and to continually match the imagination. That is their meditation. With seven drummers doing it, it is a very complicated rhythmic cycling pattern. It causes the person to go into a meditative state. Besides the drummers, the dijeridoo players were droning. In the Mandeville Auditorium [at University of California, San Diego] there are wonderful catwalks where there were seven flute players doing a telepathic meditation. You would hear flute sounds, calls and responses and chords that came from that telepathic meditation.

The audience at *Crow Two* described feeling overwhelmed at one point in the performance when a moment of utter silence was pierced by screams from crow *heyokas* (sacred clowns). Pauline Oliveros (1932–2016) had constructed a theater piece that centered on female crow figures, representing a matrilineage, who aligned their meditational powers with a female crow poet in the center. When the drummers and didjeridu (correct spelling for an Australian wind instrument) players fell silent, male heyokas burst in to disrupt the quiet. They were driven away by crow meditations and by a shiny mylar crow kite that enticed them off-stage.

Oliveros was experimenting with the idea of unconscious sound production. She was at the time a leading sound and performance artist whose work pushed the limits of music and theater. She was accustomed to including the whoosh of lawn sprinklers, the rasp of radio static, the blare of alarm clocks, and electronic reverberations in her pieces. In the 1970s Oliveros became interested in how the human voice could expand unconsciously in response to what was happening in a circumscribed space; how, with the aid of meditation, a give-and-take could occur spontaneously between musicians and performers wherein each adjusted their sounds in response to the others. The score for *Crow Two* was very detailed in its description of the planned space but left the rendition itself open. A scholar commented that Oliveros was using meditative sound to create a kind of utopia of awareness.[13]

Oliveros once said: "I took the crow as my totem spirit bird, because I have always been fascinated by them. … Crows more and more are heralds for me. They always herald interesting and maybe mischievous tidings."[14] She was attracted to Native American understandings. But *Crow Two* draws upon several traditions for its structure and enactment; some may call this "appropriation," for the work patently misrepresents the sources of inspiration. Oliveros was a bricoleur. She tapped into Buddhist meditation without performing it as a Buddhist. She used Australian didjeridu sounds and called the piece an opera. And, finally, she appropriated deeply sacred Sioux heyoka characters for a profane Western performance. By drawing a bird species into the focus of the piece, Oliveros hinted at pan-cultural possibilities linked to nature. She was breaking the boundaries of content in much the same way as she did with sound itself in order to fashion something new and resonant.

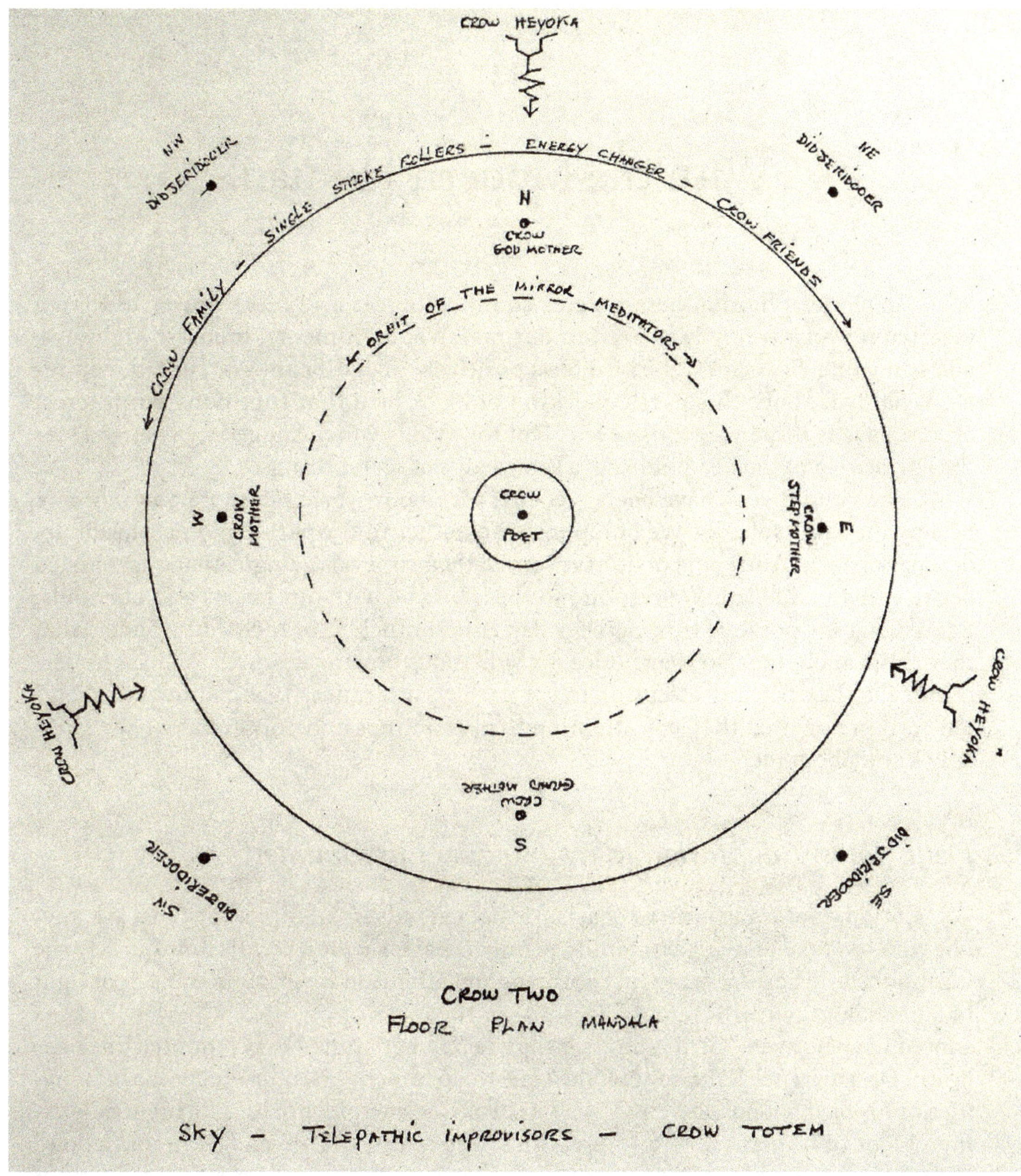


**Diagram of "Crow Two: A Ceremonial Opera." Pauline Oliveros Papers, Box 2, Folder 13. Special Collections & Archives, University of California San Diego. Permission courtesy of Special Collections and Archives, UC San Diego.**

# 12

# With the Modern Human

The modern human being seems so much bigger and more powerful than a measly crow or raven. Compared to our massive monuments, monster highways, and sprawling cityscapes, the bird feels positively insignificant. Yet, it still inspires us. What is it about this particular kind of black bird that summons our interest even today? Is it the way those eyes pivot toward us which suggests wisdom? Does the curious tip of the head indicate a knowledge of secret things?

Crows and ravens have been ever-present in our world but more and more we occupy the same spheres. We bump into each other in the parking lot, at the dump, on a park bench. And some of us have sensed the primeval dialogue that once existed between man and bird. We respond to these birds—with our hearts and our souls. I gaze up at my posse of three as they flap from branch to branch and wonder what they think about me. Do they notice my ridiculous bowing gestures as I pass underneath? Do they pay any attention to the garbled utterances I extend toward them? I'm certain now that they pay attention to my comings and goings but what else do they know about me?

## *Field Notes: Adaptation to a Changing Environment*

Corvids claim the kind of charisma that extends in all directions. Just ask anyone who has taken one in as a family member. Barb Kirpluk recalls how quickly she was hooked: "They drew me in, capturing my attention like a sudden flash of light in a dismal, gray world, much like most of the loves of my life."[1] Charles Dickens adored his pet raven "Grip" so much that he had her stuffed and mounted after her death. Dr. Lawrence Kilham took the time to document his amusement in the activities of his bird friend "Raveny."[2] And TikTok's "crowtok" broadcasts highly amusing videos of human-corvid interactions. Pet crows and ravens commonly bond with humans, but wild ones seem to know us too. They most certainly remember us.

Dr. Kevin McGowan, who spends time studying crow behavior at Cornell University, recalls: "We had one crow on [my] campus, a 4 year old, he hated my guts. He hated me so bad—he saw my car one time and followed me three blocks, and then sat in a parking lot yelling right up until I went into the building. And when I came out of the building two and a half hours later, he was still there and he started right up again. This is a bird that recognizes us out of 100 other people leaving the

**Talking to a raven. April 9, 2019. Belgrade, Serbia. This photograph captures an intimate moment of inter-species communication. Man and bird seem to derive pleasure in equal measure (AlexLinch, iStock).**

building!"[3] John Marzluff and his team of student scientists at the University of Washington in Seattle undertook extensive research on the ability of crows to recognize individual people. Wearing masks, the researchers visited crow nests and interacted with them, trapping some and tending to others. After their encounters with the masked humans, the birds responded very strongly whenever they saw the masks, even when the mask was seen in another context. When the "bad" mask was seen, the birds stirred up a commotion of scolding and dive-bombing. When the "good" mask was seen, the reaction was of calm. And they did not forget. In fact, the response increased over time. Marzulff writes: "In the five years since we trapped on campus, the number of birds scolding the caveman ['bad' mask] on a typical walk has increased threefold. And the vast majority of those who berate the Neanderthal were never even touched by him. The young learned from their parents to scold that bad man."[4] The memory of the experience crossed generations.

Crows and ravens not only recognize us, but they also seem to read our minds. This "theory of mind" was discussed briefly in Chapter 2. It advances the contention that corvids have a complex cognitive ability that allows them to imagine and consider what others (animals or humans) are thinking and what they might do. Dr. Heinrich demonstrated that ravens are aware of what their bird peers know and can anticipate their reactions. He and other researchers then experimented with the human gaze to document instances when the birds paid acute attention to our own subtle signals of intention. They had "ravens watch as a scientist gazed fixedly at a spot on the other side of a barrier. All the birds, apparently understanding that the big featherless biped knew something they did not, hopped off their perches to get a look."[5] They were taking action based on what they thought we thought.

Corvids are also cognizant of the habits of humankind; that's why they followed the sound of a battle bugle, retreated sharply at the glint of a gun, and now hop anxiously around a garbage dump. Their intelligence and curiosity intersect with an innate fear and caution. They have learned to adjust to the changes we make in their environment. Biologist Michael Westerfield points out that crows in urban settings behave differently from crows of the wooded mountainscapes. Cities are safer places and crows are more comfortable. They are bolder. In the countryside or in the fields, they are secretive and wary; someone with a rifle might be nearby.[6] Behaviors, daily routes, and migratory patterns have all shifted due to human activity as the birds innovate new responses to our mostly damaging ways. Their extremely social culture allows them to pass on their knowledge efficiently. As we expand our footprint on the world, so do they. I now see ravens, who at one time restricted themselves to the seclusion of places such as Maine, strolling outside my suburban grocery store. They're having a great time.

This, of course, has a downside. Crows are so good at adapting that they now threaten to disrupt. They have been known to blacken the skies with their numbers, descend upon crops to pull seedlings from the ground, and enter homes to pilfer eggs, sugar, or grain. They will devour native species and muscle out breeding areas. Efforts to control them mostly fail. The scarecrow is a joke; fireworks not so much. In Terre Haute, Indiana, the "Crow Patrol" set off screaming fireworks into the night sky only to watch a tornado of black feathers reassemble back into crows on a line. In Tanzania, a campaign to poison 1.2 million crows left behind an apparently small, innocuous reserve but within just a few breeding seasons the population rebounded. It's not likely we can kill or frighten them out of our lives. They're just too smart and resilient. They're too much like us. We should probably respect that. Particularly because we still have so much to learn from them. Most ornithologists will agree that we have only just scratched the surface of our understanding of their natural history and capabilities.

Contemporary authors, poets, painters, and multi-media artists also heed the corvid's call to attention. They can vouch for the place the raven and crow still have in our headspace and our body-space. Edgar Allen Poe, for instance, mused upon the raven and created one of the most recognized bird characters of all time in his poem "The Raven." Poe's bird served as a kind of fun-house mirror placed before the human predisposition for obsession and frenzy. J.R. Tolkien, Lewis Carroll, and Charles Dickens each used different aspects of ravens in their works. Portuguese writer Miguel Torga penned "Vincent the Raven" in 1941 to explore the issue of God's will. In *A Fine and Private Place* (1960), American writer Peter Beagle had a raven deliver sandwiches—not to the prophet Elijah, but to the character Jonathan Rebeck, a bankrupt pharmacist living in a cemetery. Günter Grass used crows and scarecrows as symbols in his 1963 novel *Dog Years*. And an imaginary boy called Crow became the alter ego and wisdom guide for Haruki Murakami's main character Kafka (which means "crow" in Czech) in his 2002 *Kafka on the Shore*. The crow represented all that Kafka needed—freedom and toughness—while simultaneously bearing the aura of omen and death.

Crows and ravens take on movie roles in Frank Capra's *It's a Wonderful Life* (1946), Alfred Hitchcock's *The Birds* (1963), and Alex Proyas's *The Crow* (1994). They have made appearances on TV's *The Munsters* (1964–1966) and *Game of Thrones* (2011–2019). They have been Disney cartoon characters. They even represent the Baltimore football team!

Perhaps the most famous living corvids reside at the Tower of London in England. The future of country and kingdom relies on their presence. The earliest reference connecting ravens to the Tower is an ancient Welsh tale that recounts the demise of Brân ("raven") the Blessed, King of the Britons. Brân's head is said to be buried under White Hill, upon which the Tower now stands. His strength and magic protect the country from invaders.

Other legends spin different yarns about ravens and the Tower. For instance, one claims that a flock of the talkative birds created such a ruckus around the Tower that England's royal guards came to attention just in time to stop an imminent raid by Oliver Cromwell (1599–1658), and for this reason they must be honored. The story that has been passed on most often also dates their appearance to the seventeenth century. Ravens had been particularly loathed by Londoners for their eagerness to dine on the many corpses hanging from gibbets (and on the diseased dead who lay in the streets). The city dwellers wanted the ravens destroyed and indeed they were just about exterminated. King Charles II, who reigned from 1660 to 1685, wanted them wiped out because they interfered with observatory work in the White Tower, but he did not want to tempt fate and the various prophecies that were circulating at the time so he decreed that the birds would be protected instead, as "guardians" of the Tower.

These are legends. The first concrete reference to ravens in the Tower dates to the late nineteenth century. And, to this day, it is visitors who flock to see the six to eight ravens living within the walls of the fortress. Belief holds firm that if the ravens are lost, Britain will fall. A similar practice played out across the globe at the Forbidden City in Beijing, China. During the Qing Dynasty (1636–1912) crows were fed and worshipped within the imperial grounds because they were believed to have been instrumental in saving the life of Nurhaci, the first Qing emperor (1559–1626). Though no longer cared for today, they still fly back to the Forbidden City and circle the sky in the evening.

The omnipresence of corvids, along with their peculiar charm, continues to spark the creative mind even into modern times. We take a closer look at a few of the many artists and writers who have responded to what their existence in our world implies.

## The Crow

John Clare. Written in the Northampton Asylum, England.

How peaceable it seems for lonely men
To see a crow fly in the thin blue sky
Over the woods and fealds, o'er level fen
It speaks of villages, or cottage nigh
Behind the neighboring woods—when March winds high
Tear off the branches of the huge old oak
I love to see these chimney-sweeps sail by
And hear them o'er the knarled forest croak,
Then sosh askew from the hid woodman's stroke
That in the woods their daily labours ply
I love the sooty crow, nor would provoke
Its March-day exercises of croaking joy
I love to see it sailing to and fro
While feelds and woods and waters spread below.

John Clare replied to a changed world in his work and reached backward to reconnect with the "simpler" days. Clare (1793–1864) grew up in destitute poverty among a practically illiterate farming family in rural England, suffering effects of malnutrition that left him just five feet tall. He experienced the transformation that Industrialization brought to the countryside—the factories, mines, and forest clear-cutting. His sentiments aligned with those of the great Romantic poets such as William Blake, John Keats, and William Wordsworth. He sought solace in nature.

"The Crow" is a meditation on a familiar scene—a crow flying in the sky. The sonnet's rhyming form allows it to lift up and carry the poet over the woods and fields alongside the crow. Clare makes simple observations along the way. He notices the thin blue sky, the "knarled" forest, and the woodman. He points out how the crow sails, sweeps, "sosh"s, and croaks in joy. He sees the villages, cottages, fields, woods, and waters that the crow must see. The scene is delicately exuberant. Clare includes no metaphysical or religious messages. We simply enjoy the wonder of nature, brought to us by an ordinary, sooty, crow. Yet … that first line introduces us to a lonely man. The poem starts off with a poignant, almost sad, tone. Clare sets us up as the "other," the one who can enjoy the moment but not really participate in it. This ploy heightens our sensibilities toward nature as we read on.

Clare was an intensely watchful nature-poet, known for the musicality of his poems and his animated imagery. He gives us the details of a scene rather than a narrative. Readers of "The Crow" might notice strange words, such as "sosh," and obvious misspellings such as feelds and fealds. It is believed that these elements are a nod to the oral tradition that Clare (identified as England's first working-class poet) was familiar with. In the oral tradition, the sound of a word was important, and spellings were not.

John Clare experienced brief fame with the 1820 publication of *Poems Descriptive of Rural Life and Scenery.* However, after experiencing years of mental health problems, he spent his last twenty years at the Northampton General Lunatic Asylum.

## Night Crow

Theodore Roethke. USA.

When I saw that clumsy crow
Flap from a wasted tree,
A shape in the mind rose up;
Over the gulfs of dream
Flew a tremendous bird
Further and further away
into a moonless black,
Deep in the brain, far back.

John Clare looked outward in his poems; Theodore Roethke turned inward.

Recognizing the crow as a symbol and metaphor, Roethke used the bird to guide us into the world of the human subconscious. He begins his poem "Night Crow" in a tangible place, with a clumsy bird on a wasted tree. The descriptors are specific and evocative. Things change quickly when a shape appears. It shows up in the poet's mind, but he identifies it as "the" mind. It could be our mind too. The crow takes on mythological dimensions as it balloons into a tremendous being capable of traversing space. It flies "further and further away." The distance, however, is inward into the subconscious. In a few short lines we cross "gulfs of dream"; we go from the real world to a vaguely abstract one, but just as suddenly land back where we started—in the physical part of the mind, the brain. Deposited in a moonless black landscape, alone, are we in the recesses of Roethke's unseeable inner mind or in ours?

Theodore Roethke (1908–1963) is regarded as one of the most influential poets of the twentieth century. His poems often touch upon nature, and, like Clare, he was sensitive to their aural rhythms and sound. The meter in "Night Crow" is steady and measured, almost inexorable, until it lands with a clunk: "far back." And, like Clare, his imagery is precise (even as it goes into uncharted, psychological, places): we are presented with a clumsy crow, a wasted tree, gulfs of dream, a tremendous bird, and moonless black. Roethke taps into the archetype of the crow. He couldn't have used a robin or a pigeon for this poem. This is a night crow—double darkness.

Roethke suffered a number of nervous breakdowns in his life. He brought his experiences to bear on his poems, turning a psychological liability into a poetic asset. We feel his poems. And we see the concrete elements of our world differently after reading his work. We become sensitized to how the external world can spark internal responses and how what is literal can be understood metaphorically.

## Study for *Three Muses*

Will Barnet. USA.

**Study for *Three Muses*. ca. 1980–1985. Will Barnet. Charcoal and oil stick on paper, 29.75 × 22.5 inches. In a sketch for a later painting, Will Barnet worked out a composition that includes only three features—trees, women, crows. Barnet's style was focused on the streamlining of line, shape, and color to evoke an enigmatic moment (Smithsonian American Art Museum, Gift of the artist, 1994.42.6. © 2023 Will Barnet Trust / Licensed by VAGA at Artists Rights Society [ARS], NY).**

Will Barnet (1911–2012) liked crows and used them frequently in his work. It's possible that their smooth linear shapes appealed to his artistic eye. Barnet was a prolific artist known for clean, graphic depictions of people and places. In Study for *Three Muses*, the birds hover above and around three faceless, dark-robed figures who look outward in slightly different directions. The crows almost enclose the scene, providing the only sense of movement on the canvas. They give the appearance of being "familiars" to the women. Perhaps they were summoned. The women, on the other hand, look iconic, like statues frozen just after having wrapped their robes tight to their bodies. They hold in their secrets. The scene is mysterious, mythic.

The actual oil on canvas painting that resulted from the study bears both similarities and significant differences. The composition is basically the same but in the painting the sky in the background glows with a soft white light. The crows are larger. The center crow at bottom has turned to face the muse on the right who has seated herself on the ground; one of its talons almost touches her resting hand. The women themselves are dressed in glowing white gowns and their faces resemble Elena, Barnet's wife, and his daughter and daughter-in-law. Their bodies overlap in more relaxed poses. The central figure bears a wide basket of berries. We imagine the birds have scurried down to share a morsel. The mood is serene.

Barnet tapped into two features of the crow that live side-by-side. On the one hand, its otherworldly connotations are immediate and useful. Muses, as ancient Greek goddesses of literature and the arts, might easily be associated with the crow as their messenger of inspiration. On the other hand, the birds are part of our everyday lives. They inhabit our woods and can easily be befriended.

Will Barnet was a painter and printmaker known for cryptic pictures crafted with a keen eye for formal composition. He carefully controlled space, line, and color, which resulted in pieces that are figurative and geometrical at the same time. Barnet's graceful prints and paintings speak to the classical principles of harmony and stability. He developed a personal language that is immediately recognizable.

## If You Want to Know More About the Raven

Barry H. Lopez. USA.

If you want to know more about the raven: bury yourself in the desert so that you have a commanding view of the high basalt cliffs where he lives. Let only your eyes protrude. Do not blink—the movement will alert the raven to your continued presence. Wait until a generation of ravens has passed away. Of the new generation there will be at least one bird who will find you. He will see your eyes staring up out of the desert floor. The raven is cautious, but he is thorough. He will sense your peaceful intentions. Let him have the first word. Be careful: he will tell you he knows nothing.

If you follow Barry Lopez's instructions, you will be in the desert a long time. Bring water; it just might be worth it. British writer Robert Macfarlane described Lopez as a man who believed "natural landscapes are capable of bestowing a grace upon those who pass through them. Certain landscape forms, in his vision, possess a spiritual correspondence. The stern curve of a mountain slope, a nest of wet stones on a beach, the bent trunk of a windblown tree: These abstract shapes can call out in us a goodness we might not have known we possessed."[7] Whether it was the northern tundra of his *Arctic Dreams* (1986) or the California wastelands of *Desert Notes* (1976), Barry Lopez (1945–2020) believed wildernesses can be healing. In his book *Desert Notes*, Lopez presents the desert as alive with poetry and mystery. He writes: "Here things are sharp, elemental.... Feel how still it is.... Do you hear how silent it is? This will be a comfort as you work. Do not laugh. When I first came here I laughed very loud and the sun struck me across the face and it took me a week to recover. You will only lose time by laughing."[8] Lopez wrote *Desert Notes* when he was twenty-four. He used the desert as a testing ground for self-discovery, bringing an expansive outlook to his experience. He considered when and how he could see shadows, he observed an ant, he inhabited a naked girl's body, he spoke with turtles

("They are reticent about their commitments. Each one looks like half the earth"[9]), and he made up stories—about Coyote and other people.

Lopez also pondered the merits of crows and ravens. About crows: "To begin with, the crow does nothing alone. He cannot abide silence and he is prone to stealing things, twigs and bits of straw, from the nests of his neighbors. It is a game with him. He enjoys tricks. If he cannot make up his mind the crow will take two or three wives, but this is not a game. The crow is very accommodating, and he admires compulsiveness."[10] About the raven: "Put all this to the raven: he will open his mouth as if to say something. Then he will look the other way and say nothing. Later, when you have forgotten, he will tell you he admires the crow."[11] In order to be reborn in the desert, one must be like the raven and face solitude, face oneself. Lopez's singular approach to the world opens new ways for us to grasp and use reality.

## Raven on Bowl

William Morris. USA.

With beak ajar, the raven sits on the rim of a bowl. He leans forward in amazement at his own image. We can almost hear his surprise when he sees himself. Is this Narcissus, the Greek beauty who fell in love with his own reflection in a pool of water and was unable to move? Has he turned into a raven, and into glass? Artist William Morris creates the illusion of an illusion with this amazing piece. Notice how much it looks like pottery. The cracking on the black areas reveals a terracotta underlayer. The ornamental design echoes the American Southwestern features found on vessels made by Indigenous peoples. It looks as though it could have been made a thousand years ago, or yesterday. But this is not clay. It is glass that has been pulled by hand from the furnace, treated with salts and oxides (applied to an extremely hot surface), and worked with pincers, paddles, and gloves to give the appearance of the fissures and cracks on ancient pottery. Morris works this kind of magic on many of his pieces, and many of them include crows and ravens.

Morris writes on his website:

> My work is about the symbolic meaning which is attributed to objects and/or artifacts from various cultures.... From an animist vision, gods are many and they define the world through their purpose and their character. This is how we see the physical and beyond into the metaphysical. It is these characteristics that make up a whole and complete cosmos, by adding texture, depth and mood. Crows and ravens are such gods of animism. Watching them, I am awed by the rich portrayal of their nature through intent and attitude, acting out and inventing as they go. As with many cultures, crows and ravens have been ministers to denude us to the realization that we are allied with the natural world, the real world.[12]

William Morris (b. 1957) is recognized as one of the most original and influential glass artists in America. He is renowned for his innovative glassblowing techniques. Morris grew up exploring the landscape and Native American spiritual sites around his hometown of Carmel, California, and has always been keenly aware of the bonds that link everything in the cosmos. His work evokes timelessness and modernity. It sits at a crossroads of material and imagination. When viewing Morris's pieces, we reconnect with aspects of our psyches that are eternal and ancient.

***Opposite:*** **Raven bowl. 1999. William Morris. Blown glass, 13 × 20 × 18 inches. The raven sees his own reflection in this exquisite glass bowl made by one of America's leading glass artists (Courtesy of William Morris Studio).**

## Crows

Doug Anderson. USA.

Hunch in the trees
to gossip
about God and his inexorable
experimenting,
about deer guts and fish so stupid
you could sell them air
and how out in the deserts
there's a dog called coyote
with their mind
but no wings.
Crow with Iroquois hair.
Crow with a wisecrack
for everybody,
Crow with his beak
thrust through a bun,
the paper still clinging.
Then one says something
and they all leave,
complaining
the trees are not
what they used to be.
Crow with oilslick eyes.
Crow with a knife
sheathed in a shark's fin.
Crow
in a midnight blue suit
standing in front of a judge:
Your Honor, I didn't
kill him,
just ate him
and I wasn't impressed.
Crows
clustered in the bruise light
in the bottoms
of dreams.
Crows in the red maple.
Crows keeping disrespect

respectable.
Crows teasing a stalking cat,
lifting off at the last minute,
snow shagging down
from their wings.
Crows darkening the sky,
making fun of the geese
on their way to Florida.
Crows in the roses,
beaks and thorns.
Crows feeding lizards
to their brood.
Crows lifting off road kill,
floating back down
after the car has passed.
Crow with a possum eye
speared on its beak.
Crow with a French fry.
Crows
in the chicken cages
on their way to market,
the farmer
finally gone mad.
Crows hunkered down
rumpling feathers,
announcing the cataract
of snow
over the sun.
The crows prosper.
Carrion is everywhere.
The night
that is coming
is so dark
it will feel
like fur on the eyes.
So dark suddenly
you cannot see the snow.
Thrust your hand in it.
Hear it like sand
blowing on the roof
A crow shifts his foot
and snow sifts
down from the tree.

Doug Anderson's "Crows" delivers a comprehensive catalog of the many traits attributed to the crow. This humorously insightful poem is a pleasure to read aloud. The crow here is a creature capable of wide-ranging chatter—about God and deer and fish guts. He cracks jokes and torments cats. He drives farmers mad. The trickster figure owns his reputation of disrespect and is not afraid to argue his case in front of a judge.

Anderson conveys an understanding of the real-life ways of a crow—their habit of stealing human food ("a beak thrust through a bun"), the way they fill the skies with their numbers, even what they feed their young. In the end, though, their mystery touches Anderson. In a night so dark it feels "like fur on the eyes," the crow simply shifts his foot in the snow. That's all he does. He is always there, and always with us.

Doug Anderson (b. 1943) is a poet, fiction writer, memoirist, and photographer. He grew up in Memphis, Tennessee, and served as a combat medic in the Vietnam War. Anderson received a PhD in English literature. He has published his work in numerous literary magazines and taught at several eastern colleges and universities. Of his line of work, Anderson writes:

> Poetry is the constant turning of the garden of language, of making it new in spite of using the same few thousand words over and over again. It's very hard work. It's something like musicians using the same eight notes over and over again and yet still moving and enriching us. Think about it. Poets are working for us all the time, often in the dark. Their work finds its way slowly into the language. Sometimes it comes from song lyrics, but there are even more private shamans working to save your life.[13]

## The Business of Ravens

Louis Untermeyer. USA.

What are these ravens doing in our trees,
Calling on doom and outworn prophecies?
Flying in threes

Their sinister shadow, their funereal wing
Blots the fresh color out of everything.
They do not sing,

Nor shake their throats like all the other birds;
But, in cracked monotones or broken thirds,
Their crooked words

Cowardly and contemptuous are thrown
At scarecrows who, with business of their own,
Let them alone.

In "The Business of Ravens," Louis Untermeyer shares with poet John Clare an attention to ordinary, everyday scenes. Untermeyer loved his rural home saying: "I live on an abandoned farm in Connecticut ... ever since I found my native New York unlivable as well as unlovable. ... On these green and sometimes arctic acres I cultivate whatever flowers insist on growing in spite of my neglect; delight in the accumulation of chickadees, juncos, cardinals, and the widest possible variety of songless sparrows; grow old along with three pampered cats and one spoiled cairn terrier:...."[14] In this poem, Untermeyer points out a common sight—birds in trees. He notes their blackness and cracked song; the way they gather and lurk. And he cloaks the ravens in implication. He suspects them of having negative motives. Yet, like James Thurber, Untermeyer sees the humor in a raven's aspect. For all its ominous emoting, it cannot scare a scarecrow. In his short poem, the poet has encapsulated the many connotations we assign to this bird species.

Louis Untermeyer (1885–1977) had a major influence on the appreciation of poetry in America. Although he was a high school dropout who became vice president and factory manager for his family's jewelry business, Untermeyer made his mark as an author, editor, compiler, critic, parodist, and translator of more than 100 books. He was friends with literary giants such as Ezra Pound, Arthur Miller, Sinclair Lewis, D.H. Lawrence, H.L. Mencken, and Robert Frost, and became the fourteenth poet laureate to the U.S. Library of Congress in 1961.

## Ravens

Masahisa Fukase. Japan.

***Kanazawa*. 1977. Masahisa Fukase. Photograph from *Ravens*. The sight of inky black birds crowding a treetop is familiar to just about everyone. These corvids can be interpreted in any number of ways—as haunting evil presences, as emissaries, or as friends. Fukase interpreted the ravens he encountered to be avatars for his own psychic condition. Kanazawa is a city he visited in Japan's Ishikawa Prefecture (© Masahisa Fukase Archives, courtesy of MACK).**

There is a Japanese expression *tabi-garasu* that means "traveling raven." It refers to those people who move from place to place with no permanent spot in which to settle. The phrase might well apply to artist Masahisa Fukase's (1934–2012) emotional life. Fukase led a dark and difficult existence, submitting at times to alcoholism and manic moods. With a camera always at his side, the photographer obsessively took pictures—of his large extended family, of his wife, of his cats, and of himself. The camera, in some way, shielded him.

During a particularly troubled time, Fukase obsessed on ravens. This avant-garde artist photographed them with single-minded intensity, everywhere—perched on fence posts and telephone wires, gathered in trees and around factories, even their corpses sprawled across the snow. For Fukase, the ravens seemed to be stand-ins for his own existential angst and deep loneliness. He was grieving the dissolution of his marriage and unable to make sense of his overheated condition. For ten years he captured their images (1975 to 1986), at night with a flash light as well as in the daytime. He remarked that he was drawn to the effects that flash produced on the dark sheen of the feathers and the spark that flared in their eyes.

The publication that resulted was hailed as "a high point in the photo book genre." *Karasu* [*Ravens*], first published in 1986, sold out immediately. Exhibitions of the photographs have traveled the world. One critic called the book "a hymn to heartbreak." Akira Hasegawa writes in the 2017 bi-lingual edition: "The sense of isolation and solitude exposed in this work is so potent that it is agonizing to look at, or even to avert one's gaze."[15] A grainy, raw style defines the images. Blurred shapes and moody shadows obscure the subjects. We sense an intensity of movement and life, but it feels chaotic and ill-defined. The birds truly become spectral.

Fukase wrote notes about the ravens he encountered, commenting at one time: "The raven searching in the dustbins in the rest area remains calm even when I draw close, to within about two meters. It is not scared of the camera. It looks into the lens as if in wonder, and croaks at it. I became a raven with a camera, and played following my black friend who came and went high and low in the thick fog."[16] Fukase stopped working on the project but near the end of his life he seems to have renewed his interest, noting: "…at present I'm photographing crows. I am shooting the crows with a 1,000 millimeter telephoto lens every day between 4:30 and 5:30 pm when they return to their nest, from the veranda of my office. At the same time, I am doing line drawings, and I am thinking of somehow combining the crows and the line drawings. I'll probably make black-and-white composite photographs first and add color drawing afterward."[17] Fukase did this for six months. One thousand card-sized prints that show these efforts have only recently been discovered in his archives.

Masahisa Fukase was a legendary and enigmatic artist in Japan, but little known outside the country until an exhibition hosted by the Museum of Modern Art in 1974 introduced contemporary Japanese photography to the West. His whimsical, affecting work (which included many subjects in addition to ravens and crows) was lauded. Tragically, in 1992, Fukase fell down a set of stairs at his favorite bar and remained in a coma for twenty years. During that time his oeuvre was overlooked and only after his death in 2012 was his estate slowly addressed. Thankfully, we can now appreciate his obsessions.

## Crows (乌鸦)

Cao Wenxuan; translated by Helen Wang. China.

.... And that black is truly black! As black as ink, as black as lacquer, as black as the dark night unlit by moon and stars. Yet, there is a sheen to it, and when the crow takes flight, it shimmers like satin in sunlight. Its beak, of the same hard material as the ox's horn, looks majestic. And those eyes make you feel that your earlier impression was simply unjustified: two brown-black eyes, tiny beans, shiny beads, without a hint of malevolence or hatred. On the contrary, there is an innocence, a warmth, even a degree of the affinity that one finds only in the good, kind eyes of the elderly. Now, what if we were to picture this same crow, standing, black as can be, in the glistening snow? Or, this same crow, black as can be, weaving through a shower of cherry blossom? How would we respond then? When the crow steps forward, it doesn't walk, it skips. In the past I thought that crows waddled along like ducks, but I discovered that instead of taking steps, the crow skips along with quite a sense of rhythm. And if the crow is startled when pecking at food, it will cock its head and look up with a slightly idiotic expression.

Cao Wenxuan (b. 1954) begins a delightful essay on crows by telling us that he used to fear and hate the birds. As a child, he would turn away his head and spit on the ground whenever he saw a crow. For him, and for many Chinese, its connotations are vaguely bad and unsettling. This attitude changed when he visited Tokyo, where crows are ever-present and well tolerated. Cao took a closer look. And wrote a luminous description of what he saw. In the essay, he observes their beauty ("The wings are so black and so elegant that you can appreciate why beautiful young girls in ancient tales are described as having eyebrows black as crows' wings reaching to their temples."), their habits ("their tireless racket often left me without a moment's peace in which to write. There was an electricity pole not far from my accommodation with a crow that could sustain a constant cawing from morning till night. I felt like taking a bamboo cane outside and shooing all the crows away. But I was afraid that my Japanese neighbours might see me and start saying that the Chinese treat crows badly."), and their potential for inspiration ("I returned to Beijing, and when I had settled in, I set about writing again." But for the first few days I couldn't write anything. "Why can't I write?" I asked my wife. "Because there's no crow calling from the top of the electricity pole outside," she said.)[18] Cao learned to love crows.

Cao is himself beloved as an author of children's fiction. He is often called the Hans Christian Andersen of China but his contemporary themes and his ability to paint pictures with words set him apart from Andersen. Cao grew up in extreme poverty in rural Jiangsu Province during an extended nationwide famine. His childhood experiences inform his literature. He explores the challenges and complexities of children's lives, looking at subjects such as disability, ostracism, and bigotry, as well as the ordinary difficulties in learning to grow up. Cao's work is a comfort and source of strength for younger readers.

## Ravens Are the Birds I'll Miss Most....

Louise Erdrich, USA.

Page 8 of *The Painted Drum*:

When Davan Eyke moved in, the ravens watched, but they watch everything. They are a humorous, highly intelligent bird, and knew immediately that Davan Eyke would be trouble. Therefore they dropped sticks upon the boy's roof, shat on the lintel, stole small things he left in the yard, and hid them. Pencils, coins, and once his car keys. They also laughed. The laughter of a raven is a sound unendurably human. You may know it if you have heard it in your own throat as the noise of another of Krahe's favorites, *Schadenfreude*, the joy that rises as one witnesses the pain of others. Perhaps the raven's laughter, the low rasp, sounds cynical to our ears and reminds us of the depth of our own human darkness. Of course, there is nothing human in the least about it and its source is unknowable, as are the hearts of all thing wild. "Get use to them" was all the artist [Krahe] said to Davan Eyke.

Page 15 of *The Painted Drum*:

Ravens are the birds I'll miss most when I die. If only the darkness into which we must look were composed of the black light of their limber intelligence. If only we did not have to die at all. Instead, become ravens. I've watched these birds so hard I feel their black feathers split out of my skin. To fly from one tree to another, the raven hangs itself, hawklike, on the air. I hang myself that same way in sleep, between one day and the next. When we're young, we think we are the only species worth knowing. But the more I come to know people, the better I like ravens. If I have a religious practice, it is the watching of these birds. In this house, open to a wide back field and pond, I am living within their view and territory.

Page 276, last page of *The Painted Drum*:

There must be an air bank rising and falling because the ravens are playing there. I watch as they throw themselves off a branch into the invisible stream. Over and over, they tumble into the air and fly upside down. They twist themselves upright and soar off, sink, then shoot up again over the lip of rock. Say they have eaten and are made of the insects and creatures that have lived off the dead in the raven's graveyard—then aren't they the spirits of the people, the children, the girls who sacrificed themselves, buried here? And isn't their delight a form of the consciousness we share above and below the ground and in between, where I stand, right here? As I think this, one raven veers toward me, zipping straight at my face, but I do not flinch as its wings brush through my hair. I call out my sister's name in the wildness of the moment. Then I turn and watch the raven swerve rapidly over the tops of the pines, until she plummets down the cliff again, laughs, and disappears.

The raven appears more than once in Louise Erdrich's haunting novel *The Painted Drum* (2006). It is clearly more than a bird when it shows up. The novel is a story about a sacred, healing drum—how it is first discovered, how it was originally made, and its journey of return to its people. Enmeshed in this story are many other stories—about the complications of people's lives. One of the main characters is Kurt Krahe, an artist. "Krähe" is the German word for crow.

Nature's creatures (wolves, ravens, coyotes, a dog, and a bear) affect the lives of the characters in the novel, but their appearances seem to signal more than mere presences. The raven shows up at both the beginning and at the end of the book. Its aspect sparks something within young Davan Eyke that is too much for him. He kills it heartlessly with bow and arrow. "The two humans watched as the bird simply walked away from them and entered the woods to die."[19] At the end of the novel, the raven's role has expanded. It becomes a blessing that spirits away the souls of children. The bird is the instrument for the perseverance of life. Among the Indigenous peoples of the Plains area, the raven was often characterized as a creature of metamorphosis. It symbolized transformation and was called upon by healers in medicine rituals—just as the painted drum of the novel activates a healing process.

Louise Erdrich (b. 1954) is an acclaimed author of many books of poetry and fiction. She is a member of the Turtle Mountain Band of Chippewa, an Ojibwe tribe. Like Cao Wenxuan, Erdrich draws upon her life experiences as well as the history of her people to create complex works that dig deep into the tangled dealings among humans and into the everlasting links between humans and animals.

## Solomon Gursky Was Here

Mordecai Richler. Canada.

Page 1 of *Solomon Gursky Was Here*:

One morning—during the record cold spell of 1851—a big menacing black bird, the likes of which had never been seen before, soared over the crude mill town of Magog, hard by the Vermont border, swooping low again and again. Luther Hollis brought down the bird with his Springfield.

....

dogs were pulling a long, heavily laden sled at the stern of which stood Ephraim Gursky, a small fierce hooded man cracking a whip. Ephraim pulled close to the shore and began to trudge up and down, searching the skies, an inhuman call, some sort of sad clacking noise, at once abandoned yet charged with hope, coming from the back of his throat.

....

"I say," he asked, "what happened to my raven?"

....

Ebenezer Watson took a coal-oil lamp to the window and cleared a patch of frost to keep watch.

"What did he mean, *his* raven?"

Last paragraph of *Solomon Gursky Was Here*:

Watching the Gypsy Moth [plane] climb, Moses believed that he saw it turn into a menacing black bird, the likes of which hadn't been seen over Lake Memphremagog since the record cold spell of 1851. A raven with flapping wings. A raven with an unquenchable itch to meddle and provoke things, to play tricks on the world and its creatures. He watched the bird soar higher and higher, until he lost it in the sun.

Mordecai Richler's novel begins and ends with a raven, much as Louise Erdrich's book does. In both, a raven is initially shot from the sky. But Erdrich's look at culture and landscape is lyrical and poignant, while Richler's is irreverent and boisterous. *Solomon Gursky Was Here* (first published in 1989) is a sprawling, "legendary" history of Canada from the point of view of a Jewish family's improbable arrival at the "crude mill town of Magog," to its adventures and travails (most shockingly among the Netsilik Eskimos when they are converted to Judaism) around the world, then ending with its departure in a raven-like airplane over the waters of Lake Memphremagog. In between are manic characters, hapless Arctic explorers, rumrunners, poker cheats, racketeers, scoundrels, and transgressors of all stripes.

The image of the raven is central to the novel. It dips in and out of episode after episode, carrying with it associations and symbolisms that morph and change. There is the Raven Consolidated plant in the Yellowknife gold fields, a raven skewered and harpooned onto a grave, a mysterious raven painting, a Dr. Otto Raven (Swiss financier), a radio program about the rapacious Raven Men, and Eskimo raven stories embedded in the narrative. Three characters anchor the plot: Moses Berger, who sets out to write the history of the Gursky family; Ephraim Gursky, a Jewish cantor from Minsk, Russia (who shape-shifts into a pickpocket, bogus millenarian preacher, Klondike roustabout, scam artist, and Eskimo shaman); and Solomon Gursky, Eprhaim's grandson, who sweeps through the world in his own chaotic way. Both Ephraim and Solomon are associated with raven imagery.

The Eskimo give Ephraim the name *Tulugaq*, meaning "raven." The bird will sit on his shoulder; he can talk to it. Over the course of the novel, Ephraim unapologetically takes the native women as he converts them to practices around kosher food. He merrily manipulates and misbehaves. He dons many guises and identities that seem to work even though they are completely transparent. Ephraim is the caricature of both the Wandering Jew personae (adaptable and fluid) and the Eskimo Trickster Raven (mischievous and able to take advantage of any opportunity).

Moses Berger gradually comes to understand a few things about Ephraim's grandson Solomon Gursky, too. In 1935, during a barbaric battle in China, Solomon himself tramped up and down, searching the skies.[20] A raven then appeared to lead his starving troops to safety. Reviewer Jonathan Yardly points out that Solomon is like a raven "flying through the world with lust, curiosity, and the unquenchable itch to meddle and provoke things, to play tricks on the world and its creatures, yet he is also the conscience of the Jews; as clouds form over Europe he desperately tries to open Canada's shores to refugees from the terror he sees in the making, and he insists that his religion's rituals be honored in an hour of transcendent secularism and greed."[21]

And much like the bird, the novel crosses time and space, then circles back again in a non-linear manner. Through its digressions, it addresses the myths that have emerged around Canadian national identity, the dominant settler society, and the "wildness" of the wilderness. Richler was a chronicler and a critic of everything Canadian. In his book, he boldly inserted Jewishness into the middle of a foundational Canadian narrative of the Arctic. Richler (1931–2001) was born to a second-generation Jewish family living in the working-class Jewish quarter of Montreal. The city and its people inspired much of his work during his lifetime. Richler was known as a cantankerous political and cultural commentator who had a tremendous satiric sense of humor.

# TARTAN PANTS

Jim Dine. USA.

**TARTAN PANTS. 2009. Jim Dine. Lithograph, ink on paper, on Arches vellum 400 grams, 66.5 × 52 inches). Jim Dine's two alter-egos engage in conversation (Courtesy of the Jordan Schnitzer Museum of Art, Washington State University Permanent Collection).**

Crow and raven images recur over and over in Jim Dine's artistic oeuvre. Most people don't know about the haunting heliogravure prints in his 2001 book *Birds* where crows appear in dreamlike black and white, standing mutely (sometimes menacingly) in front of the camera. They are not aware of his 2000 *Red Raven* woodcut, his 1994 etching *9 Studies for Winter Dream* (Crow), or his 1996 photogravure *Jimmy Laughing*. Instead, people think of the paintings that Dine became famous for—bright, pop art hearts and bold, graphic renderings of ordinary hardware store objects.

But Dine (b. 1935) is nostalgic about crows and he mines their connotations in works that are autobiographical, yet open to interpretation. Dine once described a childhood encounter with a crow at the Cincinnati Zoo where he heard the crow say "Hi, my name is Jimmy." He thought, "That's my name too!" It was alarming for the little boy; the adult artist has been exploring that uneasy feeling ever since. Dine looks at a crow and sees a cipher. It haunts him. He has painted and photographed the bird again and again to discover what it might have to convey.

Alter-egos have also absorbed the artist's attention, and the idea of mirroring has pervaded much of his work. Of Pinocchio (an image he worked with for a long time), Dine has said: "I saw the Walt Disney movie when I was six, and I was very frightened by it, enchanted by it. And I identify with it. I was a liar, little boys are liars."[22] This was around the same time he visited the Cincinnati Zoo.

Two alter-egos meet in TARTAN PANTS. They are large, human-sized figures (over five feet tall) presented as doubles (same nose!) talking to each other. Their postures are attentive. The boy extends his neon yellow hand toward the crow who has turned its head toward him, beak open. While not fully "pulled together," Pinocchio is solidly built, having a defined face and bright clothing. He is on his way to becoming human. The crow, on the other hand, is dripping away. It is becoming insubstantial, phantom-like. Perhaps these are two of Dine's dreams meeting in an in-between space. What could they be talking about? Only Dine knows.

Jim Dine was born an artist in Cincinnati, Ohio. Growing up, he constantly drew and painted. His parents were second-generation immigrants from Eastern Europe and practicing Jews who ran a family-owned hardware store. As his artistic career developed, Dine became associated with the emerging artists of his time like Andy Warhol, Jasper Johns, and Claes Oldenburg. His work, although often categorized as pop art, has always been personal and autobiographical. It references memories and emotions and contains metaphoric qualities. Dine has said that the Pinocchio story is for him a wonderful metaphor for the creation of art—bringing something inanimate to life. But instead of being the wooden boy, Dine is now the creator. "I am Geppetto, I am no longer Pinocchio."[23]

TARTAN PANTS is a contemporary portrayal of a deeply mythic condition. The crow and the storybook boy are as big, as present, and as significant as we are. They are alive with us. We have been brought full circle to the beginning of things—to a time when crow and raven easily dropped in and out of everyone's lives, to offer advice, give a warning, or simply play a trick on us.

# Epilogue

Magic persists in the form of a crow. Modern artists like Theodore Roethke, Will Barnet, William Morris, and Jim Dine transport us to that earlier dream-like condition when inter-species communication was possible. But even if we can no longer understand their dispatches today, a fairy dust of wonder follows crows and ravens as they go about their everyday lives. They are remarkable beings full of intrinsic mystery.

If I took Barry Lopez's advice and buried myself in the lawn with a pair of binoculars for the next twelve months, I still wouldn't totally understand what my backyard crows are all about. If I consulted all the scientific books and journals, I would only discover how much there is yet to discover. If I looked at all the art created about them, I would not be satisfied.

My crows captivate me. They fascinate with their intelligence. They offer surprises of inventiveness and feats of resilience. I respect these small creatures. And for some reason I feel like I can connect with them. Many of us, even scientists like Lawrence Kilham and Bernd Heinrich, see a crow in ourselves and something human in a raven. Dr. John Marzluff comments: "Corvids assume characteristics that were once ascribed only to humans, including self-recognition, insight, revenge, tool use, mental time travel, deceit, murder, language, play, calculated risk taking, social learning, and traditions. We are different, but by degree.... These animals inhabit our brains, and we are integrated into their brains as well."[1] So, we probably ought to spend more time with the birds, quiet our minds, hone our skills of observation, and open our hearts to what they can tell us—about ourselves, about time, and about the natural world.

If we heed the stories of old, we might apprehend even more. Mythology seems mostly silent today, but the gods are not necessarily dead, they sleep, and we can reawaken them in ourselves. At the end of the Tsimshian Raven cycle of myths, Raven's wanderings and creatings come to a conclusion. It is the end of the spirit world and, as scholar Allan Jensen writes, "Now time and space begin as man perceives them. Raven is the mediating force that brings about this transformation: while he wanders, the universe is in a state of flux between the world of spirits and the world of man ... the trickster brings two states into juxtaposition. As long as Raven exists incarnate and active, this juxtaposition exists. But when Raven's restlessness ceases, the spirit world retreats, and time begins. It is the era of man."[2]

And we now must rely on man's stories to open Raven's timeless world to us.

When Raven the creator brought the universe into existence, he brought into being species that were patterned after the spirits. The stories we have told over the ages remind us and teach us about that linkage. Through all the tellings across all the lands of the world, we come to the realization that we are godly actors—in someone else's (Raven's) larger story.

# Text Sources

## *Chapter 1*

"How the Crow Came to Be Black" p. 395. From *American Indian Myths and Legends* by Richard Erdoes and Alfonso Ortiz, copyright © 1984 by Richard Erdoes and Alfonso Ortiz. Used by permission of Pantheon Books, an imprint of the Knopf Doubleday Publishing Group, a division of Penguin Random House LLC, and The Random House Group Limited. All rights reserved.

"The Raven and The Loon" p. 90 from *Across Arctic America: Narrative of the Fifth Thule Expedition* by Knud Rasmussen, 1927, New York: G.P. Putnam's Sons.

"Wohpekumeu and Crow" p. 317 from YUROK MYTHS by Alfred L. Kroeber, copyright © 1976. Used by permission of University of California Press through Copyright Clearance Center, Inc.

"Wak and the Raven" p. 271 from *The Lure of The Honeybird: The Storytellers of Ethiopia* and https://www.ethiopianenglishreaders.com/22-stories/somalia/72-wak-and-the-raven by Elizabeth Laird, 2013, Edinburgh: Birlinn Ltd. Reproduced with permission of Birlinn Ltd through PLSclear.

"The Crow and the Crane" p. 27 from *The Bulletin* v.18, no. 897 (1897). Retrieved from https://nla.gov.au/nla.obj-677736905/view?partId=nla.obj-677746541#page/n27/mode/1up

"Why Crow Is Black and Men Find Precious Stones in The Earth" pp. 77–81 from *The Earth Is on a Fish's Back* by Natalia Maree Belting, copyright © 1965, New York: Holt, Rinehart & Winston. Permission courtesy of the Belting family.

"The Crow and The Soap" pp. 75–78 from *Once the Mullah* by Alice Geer Kelsey, copyright ©1954, London: Longmans, Green & Co. Permission courtesy of the family of Alice Geer Kelsey.

"Raven and the Fear of Growing White" by Duane Niatum from *Greenfield Review*, Vol. 9, No. 3/4, 1982. © Duane Niatum. Permission courtesy of Duane Niatum.

## *Chapter 2*

"The Beginning" pp. 384–385 from *The North Alaskan Eskimo: A Study in Ecology and Society* by Robert F. Spencer, 1959, Bureau of American Ethnology Bulletin 171, Washington, D.C.: Smithsonian Institution Bureau of American Ethnology.

"How the Earth Was Made and How Wood-Chips Became Walrus" pp. 151–154 from *Chukchee Mythology* by Waldemar Bogoras, 1910. Memoirs of the AMNH; v. 12, pt.1; Publications of the Jesup. North Pacific Expedition; v. 8, pt. 1. Leiden, E.J. Brill ltd; New York, G.E. Stechert & Co.

"Light Comes to Mankind" p. 253 from *Intellectual Culture of the Iglulik Eskimos* by Knud Rasmussen, 1929, Report of the Fifth Thule Expedition 1921–1924, Vol. VII, No.1. Copenhagen: Gyldendals Forlaostrykkeri.

"The Origin of The Raven and The Macaw" pp. 384–386 from *Outlines of Zuni Creation Myths* by Frank Hamilton Cushing, 1896, Thirteenth annual report of the Bureau of Ethnology, 1891–92, Washington, D.C.: Smithsonian Bureau of American Ethnology.

"Why the Tides Ebb and Flow" p. 201 from "Tahltan Tales" by James A. Teit, 1919, *The Journal of American Folklore* Vol. 32, No. 124.

"Doru the Birdman" pp. 182–184 from *Revisiting My Pygmy Hosts* by Paul Schebesta (G. Griffin Trans.), 1936, London: Hutchinson &. Co.

"Crow Blacker than Ever" from *Collected Poems* by Ted Hughes. Copyright © 2003 by The Estate of Ted Hughes. Reprinted by permission of Farrar, Straus and Giroux and Faber and Faber, Ltd. All Rights Reserved. All Rights Reserved.

## *Chapter 3*

"The Sun, the Moon, and Crow Crowson" pp. 318–320 from *Russian Folk-Tales* by Leonard A. Magnus, Leonard (Ed. &. Trans.), 1916, New York: E.P. Dutton & Co.

"A Living Corpse" from *Legendy I Rasskazy O Shamanach U. Yakutov, Buryat I Tungusov* by G.V. Ksenofontov, 1930 (S.N. Klemenc, Trans. 2023). Translation courtesy of S.N. Klemenc.

"The Origin of Daylight" pp. 204–205 from "Tahltan Tales" by James A. Teit, 1919, *The Journal of American Folklore* Vol. 32, No. 124.

"The Bringing of the Light by Raven" pp. 481–485 from *The Eskimo About Bering Strait* by Edward

W. Nelson, 1902, Eighteenth Annual Report of the Bureau of American Ethnology, 1896–1897, Washington, D.C.: Smithsonian Institution Bureau of American Ethnology.

"Crow That Saved the Sun" pp. 68–69 from *The Ainu of Japan* by the Rev. John Batchelor, 1892. London: The Religious Tract Society.

"The Great Archer Yi Shoots Down Nine Suns" pp. 65–69 from *Classical Chinese Myths* by Jan Walls & Yvonne Walls, copyright © 1984. Hong Kong: Joint Publishing Co. Permission courtesy of Jan Walls.

"What the Crow Said" from *Fables* by Michael Hannon, 1988. Turkey Press. Used by permission of M. Hannon.

## Chapter 4

"Let Us Do Justice to the Raven" Book 10, Chapter 60 from *The Natural History of Pliny* by Pliny the Elder (John Bostock, M.D., F.R.S., & H.T. Riley, Esq., BA., Trans.), 1855, London: Taylor and Francis.

"Conference of the Ravens" & "Raven Language" cxvii & cxix from *Legends of Iceland* by Jón Árnason (G.E.J. Powell & E. Magnøsson, Trans.), 1866, London: Longmans, Green & Co.

"Do You Understand Ravenish?" p. 91 from *Hans Anderson's Fairy Tales* by Hans Christian Anderson, 1913, New York: Constable & Co., Ltd.

"Cries of the Raven" p. 606 from *Ethnology of the Kwakiutl* by Franz Boas, 1921, Thirty-fifth Annual Report of the Bureau of American Ethnology, 1913–1914, Smithsonian Institution Bureau of American Ethnology, Washington, D.C.: Government Printing Office.

"Vāregan: The Voice of Heaven" p. 121 from *Mithraic Iconography and Ideology* by Leroy A. Campbell, 1968. Leiden: E.J. Brill. Used with permission of E.J. Brill through Copyright Clearance Center, Inc.

"Crow Goes Hunting" from *Collected Poems* by Ted Hughes. Copyright © 2003 by The Estate of Ted Hughes. Reprinted by permission of Farrar, Straus and Giroux and Faber and Faber, Ltd. All Rights Reserved. All Rights Reserved.

## Chapter 5

"Preface to the Table of Divination" pp. 33–35 from "Bird Divination among the Tibetans (Notes on Document Pelliot No. 3530, with a Study of Tibetan Phonology of the Ninth Century)" by Berthold Laufer, 1914, *T'oung Pao, 15*(1).

"The Kunbis" p. 48 from *The Fear of the Dead in Primitive Religion*, Vol. 1 by Sir James George Frazer, 1933, London: Macmillan Publishers.

"Ghost Dance Songs" pp. 982, 983, 984, 993, 1072 from *The Ghost-Dance Religion and the Sioux Outbreak of 1890* by James Mooney, 1897, Fourteenth annual report of the Bureau of Ethnology, 1892–93, Smithsonian Institution Bureau of American Ethnology. Washington, D.C.: Government Printing Office.

"The Epic of Gilgamesh" pp. 355–357 from "The Gilgamesh Narrative, Usually Called the Babylonian Nimrod Epic" by William Muss Arnolt, 1904, *Assyrian and Babylonian Literature: Selected Translations*, Robert Francis Harper (Ed.), New York: D. Appleton and Co.

Genesis: 7:17–8.17. Bible New King James Version.

"Shoot Arrows into the Geomungo-Case" from *Samguk Yusa: Legends and History of the Three Kingdoms of Ancient Korea*, David A. Mason (Trans.). Used by permission of D.A. Mason.

"Apollo's Raven" Book II: 531–565 (p. 61); 566–595 (p. 63); 596–611 (abbreviated) (p. 64) from Ovid. *The Metamorphoses*, A.S. Kline (Trans). Permission courtesy of Dr. Adam David Kline.

"Song of the Owl God" pp. 108–110 from *Songs of Gods, Songs of Humans: The Epic Tradition of the Ainu* by Donald L. Philippi, copyright © 1979. Used by permission of Princeton University Press through Copyright Clearance Center, Inc..

## Chapter 6

"The Raven Teaches the Man" pp. 155–157 from *Pirkê De Rabbi Eliezer: (The Chapters of Rabbi Eliezer the Great) According to the Text of the Manuscript Belonging to Abraham Epstein of Vienna* by Gerald Friedlander (Trans.), 1916, London: Kegan Paul, Trench, Trubner & Co.

"Crow Flying" p. 291 from "Nabaloi Law and Ritual" by C.R. Moss, 1920, *American Archeology and Ethnology* Vol. XV, No. 3. Berkeley: University of California Press.

"Hugin and Munin, Odin's Ravens" from "Óláfr hvítaskáld Þórðarson, Fragments 4" by Tarrin Wills (Ed.), 2017, in Kari Ellen Gade and Edith Marold (Eds.), *Poetry from Treatises on Poetics. Skaldic Poetry of the Scandinavian Middle Ages 3*. Turnhout: Brepols. https://skaldic.org/m.php?p-verse&i-5523. Permission courtesy of Tarrin Wills.

"The Dream of Rhonabwy" pp. 307–311 from *The Mabinogion* by Lady Charlotte Guest (Trans.), 1877, London: Bernard Quatritch.

"The Death of Cuchulainn" p. 183 from *The Mythology of the British Islands* by Charles Squire, 1905, London: Blackie and Son Ltd.

"The Tengus Rescue Tametomo." Retelling by author.

"Mortal Men" p. 238 from *Contributions to the Ethnology of the Haida* by John R. Swanton, 1905, The Jesup North Pacific Expedition, Memoirs of the AMNH Vol. V (Franz Boas, Ed.). Leiden: E.J. Brill.

"Nâs-ca'kî-yêl" p. 81 from *Tlingit Myths and Texts* by John R. Swanton, 1909, Smithsonian Institution Bureau of American Ethnology Bulletin

39. Washington, D.C.: Government Printing Office.

"Crow Makes a Woman [and Institutes Death]" p. 85 from *Eagle and Crow: An Exploration of An Australian Aboriginal Myth* by Johanna M. Blows, copyright © 1995. Reproduced by permission of Taylor & Francis Group through PLSclear.

## *Chapter 7*

"A Shaman's Magic Words" pp. 114–115 from *Intellectual Culture of the Iglulik Eskimos* by Knud Rasmussen, 1930, Copenhagen: Gyldendal.

"Word Charm" p. 120 from *Malay Magic: Being an Introduction to the Folklore and Popular Religion of the Malay Peninsula* by Walter W. Skeat, 1900, London: MacMillan and Co., Ltd.

"Midē Picture-Song #8" p. 20 from *Alcheringa* No.4, Autumn, 1972. Permission courtesy of Howard Norman.

"Raven Magic" pp. 331–333 from *Ritual and Belief in Morocco* by Edward Westermarck, 1926, London: MacMillan and Co., Ltd.

"The Raven-Stone" p. 197 from *Curiosities of Indo-European Tradition and Folklore* by Walter K. Kelly, 1863, London: Chapman and Hall.

"Come, Raven!" p. 68 from *Origin Myth of Acoma and Other Records* by Matthew W Stirling, 1942, Smithsonian Institution Bureau of American Ethnology Bulletin 135. Washington, D.C: U.S. Government Printing Office.

"The Battle of the Birds" pp. 25–27 from *Popular Tales of the West Highlands,* Vol. I. by John Francis Campbell, 1890, London: Alexander Gardner.

"A Crow's Magic" p. 82 from *Ainu Folkore: Traditions and Culture of the Vanishing Aborigines of Japan* by Carl Etter, copyright © 1949. Used by permission of National Anthropological Archives, Smithsonian Institution.

"Crow Falls in Love With His Mother-in-Law" as told by Mrs. Angela Sidney p. 67 from *Athapaskan Women: Lives and Legends* by Julie Cruikshank, 1979, National Museum of Man Mercury Series. Canadian Ethnology Service Paper No. 57. Ottawa: National Museums of Canada. Permission courtesy of Julie Cruikshank.

## *Chapter 8*

"Brotherhood" pp. 218–228 from "A Comparative Translation of the Arabic Kalīla Wa-Dimna, Chapter VI" by W.N. Brown, 1922, *Journal of the American Oriental Society, 42.*

"The Fox and the Crow" pp. 193–194 from *Æsop's Fables: A New Version by Thomas James, 1848, London: John Murray.*

"The Sycophantic Fox and The Gullible Raven" pp. 81–83 from *Fables for the Frivolous (with Apologies to La Fontaine)* by Guy Wetmore Carryl, 1899, New York & London: Harper & Brothers Publishers.

"The Fox and the Crow" and "Variations on a Theme" pp. 40–46 from *Further Fables For Our Time* by James Thurber. Copyright ©1956, renewed 1967 by Rosemary A. Thurber. Reprinted by arrangement with Rosemary A. Thurber and The Barbara Hogenson Agency, Inc. All rights reserved.

## *Chapter 9*

"Rhodian Crow Song" pp. 129–130 from *Athenaeus, Vol. IV,* translated by Charles Burton Gulick, Loeb Classical Library Volume 235, Cambridge,, Massachusetts: Harvard University Press, 1930. Loeb Classical Library ® is a registered trademark of the President and Fellows of Harvard College. Used by permission. All rights reserved.

"Saint Paul and the Raven" from CopticChurch.net. Courtesy of St. Mark's Coptic Orthodox Church, Jersey City, New Jersey.

"The Theft of Pine Nuts" pp. 256–257 from *Some Western Shoshoni Myths* by Julian H. Steward, 1944, Smithsonian Institution Bureau of American Ethnology Anthropological Papers No. 31, Bulletin 136. Smithsonian Institution: Washington, D.C.

"Why Crow Has the Right to the First Corn" p. 63 from *Myths and Legends of the New York State Iroquois* by Harriet Maxwell Converse & Arthur Caswell Parker, 1908, New York State Museum bulletin 125. Albany, New York: University of the State of New York.

"The First Corn" p. 594 from *Zuni Origin Myths* by Ruth Bunzel, 1929–1930, Forty-seventh Annual Report of the Bureau of American Ethnology. Smithsonian Institution: Washington, D.C.

"How the Kamilaroi Acquired Fire" pp. 304–305 from "Folk-Tales of the Aborigines of New South Wales (Continued)" by R.H. Mathews, 1908, *Folklore, 19*(3).

"The Raven and the Star Fruit Tree." Retelling by author.

## *Chapter 10*

"A-Ni'äne'thahi'nani'na Nisa'na" p. 972 from *The Ghost-Dance Religion and the Sioux Outbreak of* 1890 by James Mooney, 1897, Fourteenth annual report of the Bureau of Ethnology, 1892–1893. Smithsonian Institution Bureau of American Ethnology. Washington, D.C.: Government Printing Office.

"Raven Meets Man" pp. 452–453 from *The Eskimo about Bering Strait by Edward Nelson,* 1900, Eighteenth Annual Report of the Bureau of American Ethnology. Washington, D.C.: Government Printing Office.

"The Little Crow" pp. 47–51 from *When the Stones Were Soft: East African Fireside Tales* by Eleanor

B. Heady, 1968, New York: Funk & Wagnalls. Used by permission of Dianne Johnson-Feelings for The Tom Feelings Collection, LLC

"The Crow" pp. 92–94 from *The Yellow Fairy Book* by Andrew Lang (Ed.), 1897, London: Longmans, Green, and Co.

"The Story as Told by Cornix" Book II: 531–565 (p .61); 566–595 (p. 63); 596–611 (abbreviated) (p. 64) from Ovid. *The Metamorphoses (A.S. Kline, Trans.).* Permission courtesy of Dr. Adam David Kline.

"The Man Who was Changed into a Crow" pp. 278–284 from *Strange Stories from a Chinese Studio* by P'u Sung-ling (Herbert A. Giles, Trans.), 1880, London: T. De la Rue & Co.

"Tempo" by Noreen Lawlor. Permission courtesy of Noreen Lawlor.

## *Chapter 11*

"The Crow Has Given Me the Signal" p. 996 from *The Ghost-Dance Religion and the Sioux Outbreak of 1890* by James Mooney, 1897, Fourteenth annual report of the Bureau of Ethnology, 1892–1893. Smithsonian Institution Bureau of American Ethnology. Washington, D.C.: Government Printing Office.

"Raven Dance" pp. 226–227 from *Lieutenant Zagoskin's Travels in Russian America,* 1842–1844 by Laurentii A. Zagoskin (Henry N. Michael, Ed.), copyright © 1967. Permission courtesy of the Penn Museum.

"Calling the Medicine Poison" pp. 343–344 from "Indians in Overalls" by Jaime de Angulo, 1950. *The Hudson Review,* 3(3). Permission courtesy of Dr. David L. Miller.

"Angwusnasomtaqa, Crow Mother Kachina" pp. 28–29 from *Hopi Kachinas* by Edwin Earle, 1971, New York: Museum of the American Indian.

"The Raven Crown" from "Bhutan crowns young king to guide young democracy" by Simon Denyer, November 6, 2008, Reuters. Permission courtesy of Simon Denyer.

"Song of Kararat" pp. 66–68 from *Songs Of Gods, Songs of Humans: The Epic Tradition of the Ainu* by Donald L. Philippi, copyright © 1979. Used by permission of Princeton University Press through Copyright Clearance Center, Inc.

"Crow Two: A Ceremonial Opera" from Interview with Pauline Oliveros by Moira Roth, 1978, Pauline Oliveros Papers, Special Collections & Archives, University of California San Diego. Permission courtesy of Special Collections and Archives, UC San Diego.

## *Chapter 12*

"The Crow" p. 446 from *The Poems of John Clare,* Vol. II. by J.W. Tibble (Ed.), 1935, London: J.M. Dent & Sons, Ltd.

"Night Crow" copyright © 1944 by Saturday Review Association, Inc. Copyright © 1966 and renewed 1994 by Beatrice Lushington; p. 47 from *Collected Poems* by Theordore Roethke. Used by permission of Doubleday, an imprint of the Knopf Doubleday Publishing Group, a division of Penguin Random House LLC and Faber and Faber, Ltd. All rights reserved.

"If You Want to Know More About the Raven" p. 22 from *Desert Notes* by Barry Holstun Lopez, copyright © 1976. Reprinted by permission of SLL/Sterling Lord Literistic, Inc.

"Crows" pp. 53–55 from *Blues for Unemployed Secret Police* by Doug Anderson, 2000, Willimantic, Connecticut: Curbstone Press. Permission courtesy of Doug Anderson.

"The Business of Ravens" p. 306 by Louis Untermeyer from *Poetry: A Magazine of Verse* Vol. XXXI No.VI (March 1928), Harriet Monroe (Ed.). Retrieved from https://www.poetryfoundation.org/poetrymagazine/browse?contentId=17928

"Crows" by Cao Wenxuan (Helen Wang, Trans.). Permission courtesy of Helen Wang and Cao Wenxuan.

"Ravens are the birds I'll miss most…" pp. 8, 15, 276 from *The Painted Drum* by Louise Erdrich, Copyright © 2005 by Louise Erdrich. Used by permission of HarperCollins Publishers and HarperCollins Publishers Ltd.

"Solomon Gursky Was Here" p. 1, 507 from *Solomon Gursky was Here* by Mordecai Richler, 1991. London: Vintage Books.

# Chapter Notes

## *Introduction*

1. Kilham, 1989, p. 3.
2. Dickens, 1899, Chapter 6, p. 73.
3. Contreras, 1982.
4. Ingersoll, 1923, p. 154.
5. Eliade, 1964.
6. Pliny, 1940, p. 373.
7. Pollard, 1977, p. 136.
8. Campbell, 1983, p. 8.

## *Chapter 1*

1. Cao, 2015.
2. Bass, 2009, p. 56.
3. Oosten & Laugrand, 2006, p. 189.
4. Spalding, 1979, p. 78.
5. Krutak, 2014, p. 29.
6. Kroeber, 1976, p. 128.
7. *Ibid.*, p. 315.
8. Mudrooroo, 1994, pp. 35–36.
9. A'lam, 1993.
10. Goodell, 1979, p. 134.
11. *Ibid.*, p. 144.
12. White Crow of Kalasha mythology, 2022.
13. Audubon, 1870, pp. 79, 82.
14. Hyman, 1980, p. 45.
15. Clark, 1992, p. 177.

## *Chapter 2*

1. Kilham, 1989.
2. Beck, 1980.
3. Klein, (a). n.d.
4. Browne, 1996.
5. Turner, 2016, p. 50.
6. Telis, 2010.
7. Chadwick, 1999.
8. Heinrich & Bugnyar, 2007.
9. Heinrich, 1999, p. 356.
10. Emery & Clayton, 2004.
11. Spencer, 1959, p. 383.
12. Sawada, 2002, p. 3.
13. Ramsey, 1978, p. 115.
14. Hughes, 1967, p. 10.

## *Chapter 3*

1. Wallace, 1983, p. 38.
2. Van Vuren, 1984.
3. Heinrich, 1999, p. 291.
4. Turner, 2016, p. 34.
5. Kilham, 1989, p. 137.
6. Armstrong & Hermans, 2004.
7. Heinrich, 1999, p. 294.
8. Magnus, 1916, p. 341.
9. Bogoras, 1904.
10. Oosten & Laugrand, 2006, p. 194.
11. Teit, 1919, p. 213.
12. Nelson, 1902, pp. 425–426.
13. Batchelor, 1901, p. 448.
14. Chamberlain, 1919.
15. Allan, 1991, p. 36.
16. Tseng, 2011, p. 277.
17. Cooley, 2018.

## *Chapter 4*

1. Heinrich, 1999, p. 196.
2. Kilham, 1989, p. 58.
3. Marzluff & Angell, 2005, p. 199.
4. *Ibid.*, 198.
5. Heinrich, 1989, p. 21.
6. Marzluff & Angell, 2005, p. 198.
7. Brody, 1997.
8. Bhanoo, 2012.
9. Lorenz, 1952, p. 86.
10. Russell, 1994, pp. 68–69.
11. Heinrich, 1999, p. 195.
12. Kilham, 1989, p. 56.
13. Laycock, 1982, p. 28.
14. Heinrich, 1999, p. 201.
15. Hagan, 2022.
16. Pliny, 1855a, Dedication.
17. Árnason, 1864, 233.
18. Andersen, 1913, p. 97.
19. Campbell, 1968, p. 379.
20. Faas, 1971, p. 20.

## *Chapter 5*

1. Lawrence, 1986, p. 188.
2. Battiata, 2003, p. 21.

3. Heinrich, 1999.
4. Dooley, n.d.
5. Spenser, 1995.
6. Laufer, 1914, p. 46.
7. Spiritual Science Research Foundation, n.d.
8. *Ibid.*
9. Elliot, 1998, p. 222.
10. Moberly, 2000, p. 354.
11. Ilyon, 1972, p. 387.
12. Schodt, 1984.

## Chapter 6

1. Marzluff & Angell, 2005, p. 190.
2. Haupt, 2009, p. 70.
3. Laycock, 1982, p. 29.
4. Rosemary, 1995.
5. Marzluff & Angell, 2012, p. 139.
6. Zimmer, 2015.
7. Marzluff. & Angell, 2005, p. 190.
8. Lipscombe-Southwell, 2022.
9. Wallace, 1983, p. 37.
10. Marzluff, & Angell, 2005, p. 194.
11. Maududi, n.d.
12. Moss, 1920, p. 290.
13. McKinley, 2015, p. 466.
14. Ross, 2012.
15. Nordal, 2012.
16. Parker, 2010.
17. Davis, 1912, p. 355.
18. *Ibid.*
19. Stanner, 1963, p. 257.

## Chapter 7

1. Caffrey, 2001.
2. Heinrich, 1989, pp. 47, 44.
3. Bruemmer, 1984, p. 35.
4. Kilham, 1989, p. 45.
5. *Ibid.*
6. Bruemmer, 1984.
7. Heinrich, 1999, p. 266.
8. O'Grady, 2016.
9. Emery & Clayton, 2004.
10. O'Grady, 2016.
11. Heinrich, 1999, p. 262.
12. Heinrich & Bugnyar, 2007, p. 71.
13. Heinrich, 1999, p. 262.
14. *Ibid.*, p. 267.
15. Dry, 2022.
16. Rasmussen, 1929, p. 109.
17. *Ibid.*, p. 111.
18. *Ibid.*, p. 112.
19. *Ibid.*, p. 113.
20. *Ibid.*, p. 114.
21. Roufs, n.d.
22. *Ibid.*
23. Westermark, 1926, p. v.
24. *Ibid.*, p. 10.
25. Lloyd, 1854, p. 307.
26. Camillo, 1750, p. 91.
27. Stirling, 1942, p. 68.
28. Campbell, 1890, p. iii.
29. Thompson, 1990, p. 94.
30. Cruikshank, 1979, p. 58.
31. Crosman, 1993, p. 16.
32. Wyeth, n.d.
33. *Ibid.*
34. Crosman, et al., 1998, p. 129.

## Chapter 8

1. Teit, 1919, p. 208.
2. Teel, 2010.
3. Heinrich, 1999, p. 139.
4. O'Grady, 2016.
5. Kilham, 1989, p. 114.
6. Laycock, 1982, p. 28.
7. Chadwick, 1999, p. 107.
8. James, 1848, p. xvii.

## Chapter 9

1. Aronsohn, 1999, and Cammidge, 2021.
2. Brody, 1997.
3. Humane Society of the United States, n.d.
4. Wallace, 1983, p. 37.
5. Zach, 1979.
6. Kilham, 1989, p. 188.
7. Heinrich, 1999, p. 156.
8. *Ibid.*, p. 298.
9. Aronsohn, 1999.
10. Mattingly, 2014.
11. Shakespeare, *Hamlet*, Act 3, Scene 2:278.
12. Converse, 1908, p. 63.
13. *Ibid.*, p. 10.
14. Bunzel, 1932, pp. 483–488.

## Chapter 10

1. Rasmussen, 1931, p. 208.
2. Danišová, 2018.
3. Bruemmer, 1984, p. 35.
4. Heinrich, 1999, p. 64.
5. Haupt, 2009, p. 129.
6. Westerfield, 2011, pp. 38–40.
7. Nelson, 1902, p. 452.
8. Nelson, 1877–1878, p. 16.
9. Heady, 1965, p. 10.
10. Heady, 1968, p. 11.
11. Lang, 1897, p. x (preface).
12. Anderson, 1997, p. 8.
13. Jones, 2015.

## Chapter 11

1. Heinrich, 1989, p. 312.
2. Heinrich, 1999, p. 208.
3. Gilligallou & Gilligallou, 2020.
4. Regad, 2023.
5. de Angulo, 1950, pp. 340–341.
6. de Angulo & Boushey, 1975, p. 61.

7. Wall, 1989, p.162.
8. Sekaquaptewa, & Washburn, 2004, p. 471.
9. Ploeger, 2012, p. 46.
10. Everson, 2011.
11. Samuel & David, 2016.
12. Philippi, 1979, p. xi.
13. Ramirez, 2020.
14. Roth, 1978, p. 16.

## Chapter 12

1. Kirpluk, 2005, p. xiii.
2. Kilham, 1989.
3. Battiata, 2003, p. 21.
4. Marzluff & Angell, 2012, p. 187.
5. Berreby, 2005.
6. Westerfield, 2011, p. 3.
7. Macfarlane, 2005.
8. Lopez, 2021, pp. 12–13.
9. *Ibid.*, p. 29.
10. *Ibid.*, p. 19.
11. *Ibid.*, p. 20.
12. Morris, n.d.
13. Anderson, 2022.
14. "Louis Untermeyer," n.d.
15. Hasegawa, 2017, 135.
16. Kosuga, 2017, 143.
17. *Ibid.*
18. Cao, 2015.
19. Erdrich, 2005, p. 16.
20. Richler, 1991, p. 73.
21. Yardley, 1990.
22. Skobie, 2010.
23. Souter & Archino, 2023.

## Epilogue

1. Marzluff & Angell, 2012. 198
2. Jensen, 1980, 176.

# Bibliography


## *Introduction*

Campbell, J. (1983). *The Way of the Animal Powers, Vol. 1.* London: Summerfield Press.

Contreras, P. (1982, June 13). "In Defense of the Maligned Crow." *The New York Times.* https://www.nytimes.com/1982/06/13/nyregion/in-defense-of-the-maligned-crow.html.

Dickens, C. (1899). *The Works of Charles Dickens, Vol. 1: Barnaby Rudge: A Tale of the Riots of Eighty.* London: J.M. Dent & Co.

Eliade, M. (1964). *Shamanism: Archaic Techniques of Ecstasy.* Princeton University Press.

Ingersoll, E. (1923). *Birds in Legend, Fable and Folklore.* London: Longmans, Green and Co.

Kilham, L. (1989). *The American Crow and the Common Raven.* College Station: Texas A&M University Press.

Pliny the Elder. (1940). (Original work published in 77 CE). *Natural History,* Volume III. Book X. (H. Rackham, Trans.). Cambridge, MA: Harvard University Press Loeb Classical Library 353. https://www.loebclassics.com/view/pliny_elder-natural_history/1938/pb_LCL353.373.xml?mainRsKey=V7FKrS.

Pollard, J. (1977). *Birds in Greek Life and Myth.* London: Thames and Hudson.

## *Chapter 1*

A'lam, H. (1993). Crow. *Encyclopædia Iranica, VI*(4), 403–407. https://iranicaonline.org/articles/crow.

Audubon, J.J. (1870). *The Birds of America* (Vol. IV). New York: Geo. R. Lockwood and Son.

Bass, R. (2009). *Why I Came West.* New York: Houghton Mifflin Harcourt.

Belting, N.M. (1965). *The Earth Is on a Fish's Back.* New York: Holt, Rinehart & Winston.

Cao, W. (2015, September 24). Crows (H. Wang, Trans.). *Paper Republic.* https://paper-republic.org/pubs/read/crows/.

Clark, W.B. (Ed. and Trans.). (1992). *The Medieval Book of Birds: Hugh of Fouilloy's Aviarium.* Binghamton, NY: Medieval & Renaissance Texts & Studies. https://archive.org/stream/medievalbookofbi00hughuoft/medievalbookofbi00hughuoft_djvu.txt.

Clarke, P.A. (2023). *Aboriginal People and Birds in Australia: Historical and Cultural Relationships.* Melbourne: CSIRO Publishing.

The creation story. (n.d.). *Dreamtime.* https://dreamtime.net.au/creation/.

Erdoes, R., and Ortiz, A. (Eds.). (1984). *American Indian Myths and Legends.* New York: Pantheon Books.

Goodell, G. (1979). Bird lore in southwestern Iran. *Asian Folklore Studies, 38*(2), 131–153. https://doi.org/10.2307/1177687.

Hyman, S. (1980). *Edward Lear's Birds.* Secaucus, NJ: The Wellfleet Press.

Kelsey, A.G. (1954). *Once the Mullah.* London: Longmans, Green & Co.

Kroeber, A.L. (1976). *Yurok Myths.* Berkeley: University of California Press.

Krutak, L. (2014). *Tattoo Traditions of Native North America: Ancient and Contemporary Expressions of Identity.* Arnhem, Netherlands: LM Publishers.

Laird, E. (2013). *The Lure of the Honeybird: The Storytellers of Ethiopia.* Edinburgh: Birlinn.

Laird, E. (n.d.). Wak and the Raven. https://www.ethiopianenglishreaders.com/22-stories/somalia/72-wak-and-the-raven.

Mountford, C.P. (1965). *The Dreamtime: Australian Aboriginal Myths in Paintings.* Adelaide, Australia: Rigby Publications.

Mudrooroo. (1994). *Aboriginal Mythology.* London, England: Thorsons.

Nelson, E.W. (1902). The Eskimo about Bering Strait. In *Eighteenth Annual Report of the Bureau of American Ethnology, 1896–1897* (pp. 3–518). Washington, D.C.: Government Printing Office.

Niatum, D. (1982). Raven and the fear of growing white. *Greenfield Review,* 9(3/4), 79.

Oosten, J., and Laugrand, F. (2006). The bringer of light: the raven in Inuit tradition. *Polar Record, 42*(222), 187–204. https://doi.org/10.1017/S0032247406005341.

Parker, K.L. (1897, April 24). The crow and the crane: an Aboriginal legend. *The Bulletin* v.18, no. 897. p. 27. https://nla.gov.au/nla.obj-677736905/view?partId=nla.obj-677746541#page/n27/mode/1up.

Rasmussen, K. (1927). *Across Arctic America: Narrative of the Fifth Thule Expedition.* New York: G.P. Putnam's Sons.

Spalding, A. (Ed.). (1979). *Eight Inuit Myths*. Ottawa: National Museums of Canada.

White Crow of Kalasha mythology. (2022, January 20). *Indus Tales*. https://indus-tales.com/white-crow-of-kalasha-mythology/.


## Chapter 2


Beck, B.B. (1980). *Animal Tool Behavior: The Use and Manufacture of Tools by Animals*. New York: Garland STPM Press.

Bogoras, W. (1910). *Chukchee Mythology*. Leiden, Netherlands: E.J. Brill. http://hdl.handle.net/2246/28.

Browne, M.W. (1996, January 30). Second greatest toolmaker? A title crows can crow about. *The New York Times*. https://www.nytimes.com/1996/01/30/science/second-greatest-toolmaker-a-title-crows-can-crow-about.html.

Chadwick, D.H. (1999). Ravens: Legendary bird brains. *National Geographic, 195*(1), 102–115.

Cushing, F.H. (1896). Outlines of Zuñi creation myths. In *Thirteenth Annual Report of the Bureau of Ethnology 1891–1892* (pp. 321–447). Washington, D.C.: Government Printing Office.

Emery, N. J., and Clayton, N.S. (2004). The mentality of crows: Convergent evolution of intelligence in corvids and apes. *Science, 306*(5703), 1903–1907.

Heinrich, B. (1999). *Mind of the Raven*. New York: HarperCollins.

Heinrich, B., and Bugnyar, T. (2007). Just how smart are ravens? *Scientific American, 296*(4), 64–71. https://www.scientificamerican.com/article/just-how-smart-are-ravens/.

Hughes, T. (1967). *Poetry in the Making*. London, England: Faber & Faber.

Hughes, T. (1971). *Crow: From the Life and Songs of the Crow*. New York: Harper & Row.

Kilham, L. (1989). *The American Crow and the Common Raven*. College Station: Texas A&M University Press.

Klein, J. (a). (n.d.). Crow machine. https://josh.is/crow-machine/.

Klein, J. (b). (n.d.). The CrowBox. http://www.thecrowbox.com.

Krupnik, I. (2018). Waldemar Bogoras and the Chukchee: A maestro and a classical ethnography. In E. Kasten (Ed.), *Jochelson, Bogoras, and Shternberg: A Scientific Exploration of Northeastern Siberia and the Shaping of Soviet Ethnography* (pp. 131–168). Fürstenberg/Havel, Germany: Verlag der Kulturstiftung Sibirien. https://dh-north.org/siberian_studies/publications/jochbogshternkrupnik.pdf.

Ramsey, J. (1978). Crow: Or the trickster transformed. *The Massachusetts Review, 19*(1), 111–127. https://www.jstor.org/stable/25088830.

Rasmussen, K. (1929). *Intellectual Culture of the Iglulik Eskimos*. Copenhagen, Denmark: Gyldendals Forlagstrykkeri.

Sawada, M. (2002, December). *How to Make Up the Creator-god of the Efe in the Democratic Republic of Congo: A Common Mistake by Researchers with Monotheistic Background*. Paper presented at the 45th Annual Meeting of the African Studies Association, Washington, D.C. https://www.researchgate.net/publication/316470861_How_to_make_up_the_creator-god_of_the_Efe_in_the_Democratic_Republic_of_Congo_A_common_mistake_by_researchers_with_monotheistic_background.

Schebesta, P. (1936). *Revisiting My Pygmy Hosts* (G. Griffin, Trans.). London, England: Hutchinson & Co.

Spencer, R.F. (1959). The North Alaskan Eskimo: A study in ecology and society. *Bureau of American Ethnology Bulletin, 171*, 1–490. https://repository.si.edu/handle/10088/15465.

Stevenson, M.C. (1905). The Zuñi Indians: Their mythology, esoteric fraternities, and ceremonies. In *Twenty-third Annual Report of the Bureau of American Ethnology, 1901–1902* (pp. 3–608). Washington, D.C.: Government Printing Office.

Teit, J.A. (1919). Tahltan tales. *The Journal of American Folklore, 32*(124), 198–250. https://doi.org/10.2307/534980.

Telis, G. (2010, April 20). Clever crows, complex cognition. *ScienceNOW*. https://www.science.org/content/article/clever-crows-complex-cognition.

Thornton, T.F., Deur, D., and Adams, B. (2018). Raven's work in Tlingit ethno-geography. In G. Holton and T.F. Thornton (Eds.), *Language and Toponymy in Alaska and Beyond: Papers In honor of James Kari* (pp. 39–55). Honolulu: University of Hawai'i Press. http://hdl.handle.net/10125/24840.

Turner, P.S. (2016). *Crow Smarts: Inside the Brain of the World's Brightest Bird*. New York: HarperCollins.


## Chapter 3


Allan, S. (1991). *The Shape of the Turtle: Myth, Art, and Cosmos in Early China*. Stonybrook: SUNY Press.

Armstrong, R. H., and Hermans, M. (2004). Crazy corvids. In *Southeast Alaska's Natural World* (pp. 45–49). https://www.naturebob.com/sites/default/files/Crazy%20Corvids%20by%20Bob%20Armstrong%20and%20Marge%20Hermans.pdf.

Batchelor, J. (1892). *The Ainu of Japan*. London: The Religious Tract Society.

Batchelor, J. (1901). *The Ainu and Their Folk-lore*. London, England: The Religious Tract Society. https://archive.org/details/ainutheirfolklor00batcrich/page/68/mode/2up.

Bogoras, W. (1904). *The Chukchee*. Leiden, Netherlands: E.J. Brill. http://hdl.handle.net/2246/5745.

Chamberlain, B.H. (1919). *The Kojiki*. https://www.sacred-texts.com/shi/kj/index.htm.

Cooley, R. (2018, November 15). Cal Poly showcases local poet Michael Hannon in Pilgrim's Process. *New Times*. https://www.newtimesslo.com/sanluisobispo/cal-poly-showcases-local-poet-michael-hannon-in-pilgrims-process/Content?oid=6933391.

Hannon, M. (2011). What crow said. In E. Osborn and C. Anderson (Eds.), *A Bird Black as the Sun: California Poets on Crows & Ravens* (p. 142). Santa Barbara, CA: Green Poet Press.

Heinrich, B. (1999). *Mind of the Raven*. New York: HarperCollins.

Kilham, L. (1989). *The American Crow and the Common Raven*. College Station: Texas A&M University Press.

Ksenofontov, G.V. (1955). (Original work published in 1930). *Legendy I Rasskazy O Shamanach U. Yakutov, Buryat I Tungusov*. In A. Friedrich and G. Buddress (Eds. and Trans.), *Schamanengeschichten Aus Sibirien* (pp. 210–213). Munich, Germany: Otto Wilhelm Barth Verlag.

Magnus, L.A. (Ed. And Trans.). (1916). *Russian Folk-tales*. New York: E.P. Dutton & Co.

Mair, V.H. (Ed.). (1994). *The Columbia Anthology of Traditional Chinese Literature*. New York: Columbia University Press.

Miller, D. (Ed.). (1942). *Americans, 1942: Artists from 9 States [Exhibition Catalogue]*. New York: Museum of Modern Art.

Nelson, E.W. (1902). The Eskimo about Bering Strait. In *Eighteenth Annual Report of the Bureau of American Ethnology, 1896–1897* (pp. 3–518). Washington, D.C.: Government Printing Office.

Oosten, J., and Laugrand, F. (2006). The bringer of light: the raven in Inuit tradition. *Polar Record*, *42*(222), 187–204. https://doi:10.1017/S0032247406005341.

Teit, J.A. (1919). Tahltan tales. *The Journal of American Folklore*, *32*(124), 198–250. https://doi.org/10.2307/534980.

Tseng, L.L. (2011). *Picturing Heaven in Early China*. Cambridge, MA: Harvard University Asia Center.

Turner, P.S. (2016). *Crow Smarts: Inside the Brain of the World's Brightest Bird*. New York: HarperCollins.

Van Vuren, D. (1984). Aerobatic rolls by ravens on Santa Cruz Island, California. *The Auk*, *101*(3), 620–621.

Wallace, D.R. (1983). Ravens. *Blair & Ketchum's Country Journal*, *10*(3), 36–39.

Walls, J., and Walls, Y. (1984). *Classical Chinese Myths*. Hong Kong: Joint Publishing Co.

Yang, L., and An, D. (2005). *Handbook of Chinese Mythology*. Santa Barbara, CA: ABC-CLIO.

## Chapter 4

Andersen, H.C. (1913). *Hans Andersen's Fairy Tales*. New York: Constable & Co. Ltd.

Árnason, J. (1864). *Icelandic Legends* (G.E.J. Powell and E. Magnøsson, Trans.). London: Richard Bentley.

Árnason, J. (1866). *Legends of Iceland* (G.E.J. Powell and E. Magnøsson, Trans.). London: Longmans, Green & Co.

Bhanoo, S.N. (2012, April 23). Ravens can recognize old friends, and foes, too. *The New York Times*.

Boas, F. (1921). Ethnology of the Kwakiutl. In *Thirty-fifth Annual Report of the Bureau of American Ethnology, 1913–1914* (pp. 43–794). Washington, D.C.: Government Printing Office. https://repository.si.edu/handle/10088/91755.

Brody, J.E. (1997, May 27). The too-common crow, too close for comfort. *The New York Times*.

Campbell, L.A. (1968). *Mithraic Iconography and Ideology*. Leiden, Netherlands: E.J. Brill.

Faas, E. (1971). Ted Hughes and crow: An interview with Ted Hughes. *London Magazine*, *10*(10), 5–20.

Hagan, C. (2022, July 13). Kawanabe Kyosai, the Japanese printmaker who pioneered manga, finally gets his due. *ARTnews*. https://www.artnews.com/art-news/artists/kawanabe-kyosai-who-was-he-royal-academy-1234633991/.

Heinrich, B. (1989). *Ravens in Winter*. New York: Summit Books.

Heinrich, B. (1999). *Mind of the Raven*. New York: HarperCollins.

Hughes, T. (1971). *Crow: From the Life and Songs of the Crow*. New York: Harper & Row.

Hunt, M. (Trans. & Ed.). (1884). *Grimm's Household Tales* Vol. I & II. London: George Bell and Sons.

Kilham, L. (1989). *The American Crow and the Common Raven*. College Station: Texas A&M University Press.

Laycock, G. (1982). King of passerines. *Audubon*, *84*(6), 26–29.

Lorenz, K.Z. (1952). *King Solomon's Ring*. New York: Thomas Y. Crowell Company.

Marzluff, J., and Angell, T. (2005). *In the Company of Crows and Ravens*. New Haven, CT: Yale University Press.

Pliny the Elder. (1855a). (Original work published in 77 CE). *The Natural History of Pliny: Dedication* (J. Bostock and H.T. Riley, Trans.). London: Taylor and Francis. http://data.perseus.org/citations/urn:cts:latinLit:phi0978.phi001.perseus-eng1:1.dedication.

Pliny the Elder. (1855b). (Original work published in 77 CE). *The Natural History of Pliny: Book 10, Chapter 60* (J. Bostock and H.T. Riley, Trans.). London: Taylor and Francis. http://data.perseus.org/citations/urn:cts:latinLit:phi0978.phi001.perseus-eng1:10.60.

Russell, C. (1994). *Spirit Bear*. Toronto, Canada: Key Porter Books.

## Chapter 5

Arnolt, W.M. (1904). The Gilgamesh narrative, usually called the Babylonian Nimrod epic. In R.F. Harper (Ed.), *Assyrian and Babylonian*

*Literature: Selected Translations* (pp. 324–368). New York: D. Appleton and Co.

Battiata, M. (2003, August 3). Crows: A murder mystery. *The Washington Post Magazine.*

Dooley, J. (n.d.). Poughkeepsie's massive crow roost. *Hudson Valley Viewfinder Magazine.* https://www.scenichudson.org/viewfinder/poughkeepsies-massive-crow-roost/.

Elliot, M.A. (1998). Ethnography, reform, and the problem of the real: James Mooney's "Ghost-dance Religion." *American Quarterly, 50*(2), 201–233. https://www.jstor.org/stable/30041613.

Frazer, J.G. (1933). *The Fear of the Dead in Primitive Religion* (Vol. 1). London, England: Macmillan.

Heinrich, B. (1999). *Mind of the Raven.* New York: HarperCollins.

Ilyon. (1972). *Samguk Yusa: Legends and History of the Three Kingdoms of Ancient Korea* (H. Tae-Hung and G.K. Mintz, Trans.). Seoul, South Korea: Yonsei University Press.

Lang, B. (1985). Non-Semitic deluge stories and the book of Genesis: A bibliographical and critical survey. *Anthropos, 80*(4/6), 605–616. http://www.jstor.org/stable/40461062.

Laufer, B. (1914). Bird divination among the Tibetans: Notes on Document Pelliot No. 3530, with a study of Tibetan phonology of the ninth century. *T'oung Pao, 15*(1), 1–110. http://www.jstor.org/stable/4526388.

Lawrence, R.D. (1986). *In Praise of Wolves.* New York: Henry Holt.

Marcus, D. (2002). The mission of the raven (Gen. 8:7). *Journal of the Ancient Near Eastern Society, 29*(1), 71–80. https://janes.scholasticahq.com/article/2443-the-mission-of-the-raven-gen-8-7.

Mason, D.A. (Trans.). (2023). "Shoot Arrows Into the Geomungo-case…" [Personal communication].

Moberly, R.W.L. (2000). Why did Noah send out a raven? *Vetus Testamentum, 50*(3), 345–356. http://www.jstor.org/stable/1585294.

Mooney, J. (1897). The Ghost-dance religion and the Sioux outbreak of 1890. In *Fourteenth Annual Report of the Bureau of Ethnology, 1892–1893* (pp. 641–1110). Washington, D.C.: Government Printing Office.

Ovid. (2000). (Original work published in 8 CE). *The Metamorphoses* (A. S. Kline, Trans.). https://www.poetryintranslation.com/PITBR/Latin/Metamorph2.php#anchor_Toc64106124.

Philippi, D.L. (1979). *Songs of Gods, Songs of Humans: The Epic Tradition of the Ainu.* Princeton University Press.

Schodt, F.L. (1984). Interview with Donald L. Philippi. http://www.jai2.com/dlpivu1.htm.

Spenser, E. (1995). (Original work published in 1590 CE). *The Faerie Queene.* https://www.luminarium.org/renascence-editions/queene2.html.

Spiritual Science Research Foundation. (n.d.). Significance of feeding crows during the Shraddha ritual in Pitrupaksha. https://www.spiritualresearchfoundation.org/spiritual-problems/ancestral-spirits/feeding-crows-ancestors-shraddha-pitrupaksha/.

## *Chapter 6*

Anderson, R.B. (Trans.). (1901). (Original work published ca. 1220 CE). *The Younger Edda.* Chicago: Scott, Foresman and Company.

Blows, J.M. (1995). *Eagle and Crow: An Exploration of an Australian Aboriginal Myth.* New York: Garland Publishing.

Clarke, P.A. (2016). Birds as Totemic Beings and Creators in the Lower Murray, South Australia. *Journal of Ethnobiology, 36*(2): 277–293. http://dx.doi.org/10.2993/0278-0771-36.2.277.

ibid. (2016). Birds and the Spirit World of the Lower Murray, South Australia. *Journal of Ethnobiology 36*(4): 746–764. http://dx.doi.org/10.2993/0278-0771-36.4.746.

Davis, F.H. (1912). *Myths & Legends of Japan.* London: George G. Harrap & Company.

Friedlander, G. (Trans.). (1916). (Original work published ca. 800 CE). *Pirḳê De Rabbi Eliezer: The Chapters of Rabbi Eliezer the Great According to the Text of the Manuscript Belonging to Abraham Epstein of Vienna.* London: Kegan Paul, Trench, Trubner & Co.

Guest, C. (Trans.). (1877). (Original work published ca. 1100 CE). *The Mabinogion.* London: Bernard Quatritch. https://www.gutenberg.org/files/5160/5160-h/5160-h.htm.

Haupt, L.L. (2009). *Crow Planet: Essential Wisdom from the Urban Wilderness.* New York: Little, Brown and Company.

Laycock, G. (1982). King of passerines. *Audubon, 84*(6), 26–29.

Lipscombe-Southwell, A. (2022, May 20). As the crow dies: The strange world of bird funerals. *BBC Science Focus Magazine.* https://www.sciencefocus.com/nature/raven-crow-funerals-intelligence/.

Marzluff, J., and Angell, T. (2005). *In the Company of Crows and Ravens.* New Haven, CT: Yale University Press.

Marzluff, J., and Angell, T. (2012). *Gifts of the Crow: How Perception, Emotion, and Thought Allow Smart Birds to Behave Like Humans.* New York: Free Press.

Maududi, A.A. (Trans.). (n.d.). *Qur'an: Surah 5:* https://myislam.org/surah-maidah/ayat-31/.

McKinley, R. (2015). Human and proud of it!: A structural treatment of headhunting rites and the social definition of enemies. HAU: Journal of Ethnographic Theory, 5(2), 443–483. https://doi.org/10.14318/hau5.2.031.

Moss, C.R. (1920). *Nabaloi Law and Ritual.* Berkeley: University of California Press.

Nordal, G. (2012). Poetry and society: The circumstances of skaldic production. In D. Whaley (Ed.), *Poetry from the Kings' Sagas 1: From Mythical Times to C. 1035* (pp. xc–xciii). Turnhout, Belgium: Brepols.

Parker, W. (2010, August). The dream of Rhonabwy. https://www.mabinogion.info/rhonabwy.htm.

Rosemary, K. (1995, December 3). Guile on the wing: The black-billed magpie, artful and aggressive, has defied human persecution to thrive on wooded streamsides and sagebrush grasslands east of the Cascades. *The Spokesman-Review*. https://www.spokesman.com/stories/1995/dec/03/guile-on-the-wing-the-black-billed-magpie-artful/.

Ross, M.C. (2012). What is skaldic poetry of the Scandinavian Middle Ages? In D. Whaley (Ed.), *Poetry from the Kings' Sagas 1: From Mythical Times to C. 1035* (pp. xiii–xviii). Turnhout, Belgium: Brepols.

Sayers, W. (2022). The mythological Norse ravens Huginn and Muninn: Interrogators of the newly slain. *Studia Linguistica Universitatis Iagellonicae Cracoviensis, 139*(2), 143–155. https://doi.org/10.4467/20834624SL.22.008.15632.

Squire, C. (1905). *The Mythology of the British Islands*. London: Blackie and Son Ltd.

Stanner, W.E.H. (1963). On Aboriginal religion: Cosmos and society made correlative. *Oceania, 33*(4), 239–273. http://www.jstor.org/stable/40329401.

Swanton, J.R. (1905). *Contributions to the Ethnology of the Haida*. The Jesup North Pacific Expedition, Memoirs of the AMNH Vol. V (Franz Boas, Ed.). Leiden, Netherlands: E.J. Brill. http://hdl.handle.net/2246/5742.

Swanton, J.R. (1909). *Tlingit Myths and Texts*. Smithsonian Institution Bureau of American Ethnology Bulletin 39. Washington, D.C.: Government Printing Office.

Wallace, D.R. (1983). Ravens. *Blair & Ketchum's Country Journal, 10*(3), 36–39.

Wills, T. (Ed.). (2017). Fragments (Ólhv Frag). In K.E. Gade and E.E. Marold (Eds.), *Poetry from Treatises on Poetics* (pp. 302–310). Turnhout, Belgium: Brepols. https://skaldic.org/m.php?p=wordtextlp&i=97357.

Zimmer, C. (2015, October 1). Crows may learn lessons from death. *The New York Times*.

## Chapter 7

Bruemmer, F. (1984). Ravens: Smartest birds in the world. *International Wildlife, 14*(5), 33–35.

Caffrey, C. (2001). Catching crows. *North American Bird Bander, 26*(4), 137–145. http://www.caroleecaffrey.com/uploads/6/8/1/9/68196877/catching_crows_pdf.pdf.

Camillo, L. (1750). *The Mirror of Stones*. London, England: J. Freeman. https://archive.org/details/mirrorofstonesin00leon/page/n5/mode/2up.

Campbell, J.F. (1890). *Popular Tales of the West Highlands* (Vol. I). London, England: Alexander Gardner.

Crosman, C. (1993). *Jamie Wyeth: Islands* [Exhibition catalogue]. Rockland, ME: Farnsworth Art Museum.

Crosman, C., et al. (1998). *Wondrous Strange: The Wyeth Tradition*. Boston: Little, Brown and Co.

Cruikshank, J. (1979). *Athapaskan Women: Lives and Legends*. Ottawa: National Museums of Canada.

Dry, J. (2022, January 28). The singular Kathryn Hunter goes from crone to crow in "The Tragedy of Macbeth." *IndieWire*. https://www.indiewire.com/2022/01/tragedy-of-macbeth-kathryn-hunter-witches-shakespeare-1234694872/.

Emery, N.J., and Clayton, N.S. (2004). The mentality of crows: Convergent evolution of intelligence in corvids and apes. *Science, 306*(5703), 1903–1907.

Etter, C. (1949). *Ainu Folklore: Traditions and Culture of the Vanishing Aborigines of Japan*. New York: Wilcox & Follett.

Heinrich, B. (1989). The ravens' feast. *Natural History, 2/89*, 44–51.

Heinrich, B. (1999). *Mind of the Raven*. New York: HarperCollins.

Heinrich, B., and Bugnyar, T. (2007). Just how smart are ravens? *Scientific American, 296*(4), 64–71. https://www.scientificamerican.com/article/just-how-smart-are-ravens/.

Kelly, W.K. (1863). *Curiosities of Indo-European Tradition and Folklore*. London: Chapman and Hall.

Kilham, L. (1989). *The American Crow and the Common Raven*. College Station: Texas A&M University Press.

Lloyd, L. (1854). *Scandinavian Adventures* (Vol. II). London: Richard Bentley.

O'Grady, C. (2016, January 28). Crows: The tail-pulling, food-stealing bird prodigies. *Ars Technica*. https://arstechnica.com/science/2016/01/crows-the-tail-pulling-food-stealing-bird-prodigies/.

Norman, H.A. (1972). Ojibwa picture songs. *Alcheringa, 4*, 28–30.

Rasmussen, K. (1929). *Intellectual Culture of the Iglulik Eskimos*. Copenhagen, Denmark: Gyldendals Forlagstrykkeri.

Roufs, T.G. (Ed.). (n.d.). *When Everybody Called Me Gabe-bines, 'Forever-Flying-Bird': Teachings from Paul Buffalo*. https://www.d.umn.edu/cla/faculty/troufs/Buffalo/PB30.html.

Skeat, W.W. (1900). *Malay Magic: Being an Introduction to the Folklore and Popular Religion of the Malay Peninsula*. London: Macmillan.

Stirling, M.W. (1942). *Origin Myth of Acoma and Other Records*. Washington, D.C.: Government Printing Office.

Thompson, F. (1990). John Francis Campbell (1821–1885). *Folklore, 101*(1), 88–96. http://www.jstor.org/stable/1259886.

Westermarck, E. (1926). *Ritual and Belief in Morocco*. London, England: Macmillan.

Wyeth, J. (n.d.). *The Raven* [Image text]. Brandywine Museum of Art, Chadds Ford, PA. https://collections.brandywine.org/objects/5520/the-raven?ctx=a3ba64be-fb4f-49c8-804e-09d80b184ac3&idx=292.

## Chapter 8


Brown, W.N. (1922). A comparative translation of the Arabic Kalīla wa-Dimna, chapter VI. *Journal of the American Oriental Society, 42*, 215–250. https://doi.org/10.2307/593637.

Carryl, G.W. (1899). *Fables for the Frivolous (with Apologies to La Fontaine)*. New York: Harper & Brothers.

Chadwick, D.H. (1999). Ravens: Legendary bird brains. *National Geographic, 195*(1), 102–115.

Heinrich, B. (1999). *Mind of the Raven*. New York: HarperCollins.

*James, T. (1848). Æsop's Fables: A New Version. London, England: John Murray.* https://archive.org/details/sopsfablesnewv00jame/page/n5/mode/2up.

Kilham, L. (1989). *The American Crow and the Common Raven*. College Station: Texas A&M University Press.

Laycock, G. (1982). King of passerines. *Audubon, 84*(6), 26–29.

O'Grady, C. (2016, January 28). Crows: The tail-pulling, food-stealing bird prodigies. *Ars Technica*. https://arstechnica.com/science/2016/01/crows-the-tail-pulling-food-stealing-bird-prodigies/.

Teel, J. (2010, July 17). Ravens and grizzly bears—unique relationships [Blog post]. *The Grizzly Bear Blog*. https://grizzlybearblog.wordpress.com/2010/07/17/ravens-and-grizzly-bears-unique-relationships/.

Teit, J.A. (1919). Tahltan tales. *The Journal of American Folklore, 32*(124), 198–250. https://doi.org/10.2307/534980.

Thurber, J.G. (1956). *Further Fables of Our Time*. New York: Simon & Schuster.


## Chapter 9


Aronsohn, B. (1999). Crow hunting history. *Crow Busters*. http://www.crowbusters.com/begart14.html.

Brody, J.E. (1997, May 27). The too-common crow, too close for comfort. *The New York Times*.

Bunzel, R. (1932). Zuñi origin myths. In *Forty-seventh Annual Report of the Bureau of American Ethnology, 1929–1930* (pp. 545–609). Washington, D.C.: Government Printing Office. https://archive.org/details/annualreportofbu47smithso/page/594/mode/2up.

Cahir, F., Clark, A., & Clarke, P. (2018). *Aboriginal Biocultural Knowledge in South-eastern Australia: Perspectives of Early Colonists*. Melbourne: CSIRO Publishing.

Cammidge, J.R. (2021, April 7). *History of the Crow* [Blog post]. https://johnrcammidge.com/history-of-crows/.

Converse, H.M. (1908). *Myths and Legends of the New York State Iroquois* (A.C. Parker, Ed.). Albany: University of the State of New York.

The departure of the Great Saint Anba Paul, the First Hermit. (n.d.). *Synaxarium*. https://www.copticchurch.net/synaxarium/6_2.html#1.

Gulick, C.B. (Trans.). (1930). *Athenaeus: The Deipnosophists* (Vol. IV, Book VIII). Cambridge, MA: Harvard University Press.

Heinrich, B. (1999). *Mind of the Raven*. New York: HarperCollins.

Humane Society of the United States. (n.d.). *What to Do About Crows*. https://www.humanesociety.org/resources/what-do-about-crows.

Kilham, L. (1989). *The American Crow and the Common Raven*. College Station: Texas A&M University Press.

Mathews, R.H. (1908). Folk-tales of the Aborigines of New South Wales. *Folklore, 19*(3), 303–308. http://www.jstor.org/stable/1254515.

Mattingley, C. (2014). *Seen but Not Heard: Lilian Medland's Birds*. Canberra: National Library of Australia.

Shakespeare, W. (1599–1601). *The Tragedy Of Hamlet, Prince of Denmark*. Folger Shakespeare Library. https://folger-main-site-assets.s3.amazonaws.com/uploads/2022/11/hamlet_PDF_FolgerShakespeare.pdf.

Steward, J.H. (1944). Some western Shoshoni myths. *Bureau of American Ethnology Bulletin, 136*(31), 249–299. Washington, D.C.: Government Printing Office.

Wallace, D.R. (1983). Ravens. *Blair & Ketchum's Country Journal, 10*(3), 36–39.

Wills, G. (1970). Phoenix of Colophon's ΚΟΡΩΝΙΣΜΑ. *The Classical Quarterly, 20*(1), 112–118. http://www.jstor.org/stable/637514.

Zach, R. (1979). Shell dropping: Decision-making and optimal foraging in northwestern crows. *Behaviour, 68*(1/2), 106–117. http://www.jstor.org/stable/4533944.


## Chapter 10


Anderson, W.S. (1997). *Ovid's Metamorphoses: Books 1–5*. Norman: University of Oklahoma Press.

Bruemmer, F. (1984). Ravens: Smartest birds in the world. *International Wildlife, 14*(5), 33–35.

Danišová, N. (2018). Animal transformation as a deserved punishment in archnarratives. *Ars Aeterna, 10*(2), 18–31. https://doi.org/10.1515/aa-2018-0009.

Haupt, L.L. (2009). *Crow Planet: Essential Wisdom from the Urban Wilderness*. New York: Little, Brown and Company.

Heady, E.B. (1965). *Jambo, Sungura: Tales from East Africa*. New York: W.W. Norton.

Heady, E.B. (1968). *When the Stones Were Soft: East African Fireside Tales*. New York: Funk & Wagnalls.

Heinrich, B. (1999). *Mind of the Raven*. New York: HarperCollins.

Jones, J. (2015, December 9). *Jorge Luis Borges Picks 33 of His Favorite Books to Start His Famous*

*Library of Babel*. https://www.openculture.com/2015/12/jorge-luis-borges-picks-33-of-his-favorite-books-to-start-his-famous-library-of-babel.html.

Lang, A. (Ed.). (1897). *The Yellow Fairy Book*. London: Longmans, Green, & Co.

Lawlor, N. (2011). Tempo. In E. Osborn and C. Anderson (Eds.), *A Bird Black as the Sun: California Poets on Crows & Ravens (p. 45)*. Santa Barbara, CA: Green Poet Press.

Mooney, J. (1897). The Ghost-dance religion and the Sioux outbreak of 1890. In *Fourteenth Annual Report of the Bureau of Ethnology, 1892–1893* (pp. 641–1110). Washington, D.C.: Government Printing Office.

Nelson, E.W. (1877–1878). *Alaska, May 26, 1877–May 13, 1878*. Smithsonian Field Book Project. https://transcription.si.edu/view/26186/SIA-SIA2017-084057-000001.

Nelson, E.W. (1902). The Eskimo about Bering Strait. In *Eighteenth Annual Report of the Bureau of American Ethnology, 1896–1897* (pp. 3–518). Washington, D.C.: Government Printing Office.

Overholt, T.W. (1974). The Ghost Dance of 1890 and the nature of the prophetic process. *Ethnohistory, 21*(1), 37–63. https://doi.org/10.2307/481129.

Ovid. (2000). (Original work published in 8 CE). *The Metamorphoses* (A. S. Kline, Trans.). https://www.poetryintranslation.com/PITBR/Latin/Metamorph2.php#anchor_Toc64106124.

P'u, S. (1880). *Strange Stories from a Chinese Studio* (H.A. Giles, Trans.). London: T. De La Rue & Co.

Rasmussen, K. (1931). *The Netsilik Eskimo: Social Life and Spiritual Culture*. Copenhagen, Denmark: Gyldendals Forlagstrykkeri.

Westerfield, M. (2011). *The Language of Crows*. Willimantic, CT: Ashford Press.

## Chapter 11

de Angulo, J. (1950). Indians in overalls. *The Hudson Review, 3*(3), 327–377. https://doi.org/10.2307/3847452.

de Angulo, J., and Boushey, A. (1975). The Achumawi life-force. *The Journal of California Anthropology, 2*(1), 60–63. http://www.jstor.org/stable/27824810.

Denyer, S. (2008, November 6). Bhutan crowns young king to guide young democracy. *Reuters*. https://www.reuters.com/article/us-bhutan-coronation/bhutan-crowns-young-king-to-guide-young-democracy-idUSTRE4A51SH20081106.

Earle, E. (1971). *Hopi Kachinas*. New York: Museum of the American Indian.

Everson, G. (2011). Edward Kennard, the Federal Writers' Project, and public archaeology. *The SAA Archaeological Record, 11*(3), 34–37. http://onlinedigeditions.com/publication/?i=70732&article_id=738654&view=articleBrowser.

Gilligallou, B., and Gilligallou, L. (2020, January 16). The unmistakable courtship displays of the raven [Blog post]. *Gilligallou Bird*. https://gilligalloubird.com/blogs/blog/the-unmistakable-courtship-displays-of-the-common-raven.

Heinrich, B. (1989). *Ravens in Winter*. New York: Summit Books.

Heinrich, B. (1999). *Mind of the Raven*. New York: HarperCollins.

Mooney, J. (1897). The Ghost-dance religion and the Sioux outbreak of 1890. In *Fourteenth Annual Report of the Bureau of Ethnology, 1892–1893* (pp. 641–1110). Washington, D.C.: Government Printing Office.

Norbu, P., Wangchuk, R., and Dema, K. (2008). Coronation 2008: Celebration days. *Kuensel: Bhutan's National Newspaper*. https://www.raonline.ch/pages/bt/rfam/bt_gyalpo050603c.html.

Oliveros, P. (1975). *Crow Two*. Pauline Oliveros Papers, Special Collections & Archives, UC San Diego.

Philippi, D.L. (1979). *Songs of Gods, Songs of Humans: The Epic Tradition of the Ainu*. Princeton, NJ: Princeton University Press.

Ploeger, A.D. (2012). *The Hopi Katsina Art and Ritual: Preserving a People of Peace* (Master's thesis). University of Wisconsin–Superior. https://minds.wisconsin.edu/bitstream/handle/1793/61578/MA%20Art%20History%20Thesis_Anndrea%20Ploeger.pdf?sequence=1&isAllowed=y.

Ramirez, E. (2020). Queer theory and third-wave feminism in Pauline Oliveros's meditative works (Master's thesis). University of Arizona, Tucson, AZ. https://www.proquest.com/openview/19418c910bcf0257399e1848791112c9/.

Regad, T. (2023, March 22). Crow mating behavior [Blog post]. *Animal Behavior Corner*. https://animalbehaviorcorner.com/crow-mating-behavior/.

Roth, M. (1978). Interview with Pauline Oliveros. https://library.ucsd.edu/dc/object/bb9153146j/_1.pdf.

Samuel, G., and David, A.R. (2016). The multiple meanings and uses of Tibetan ritual dance: "Cham" in context. *Journal of Ritual Studies, 30*(1),7–24. https://www.jstor.org/stable/44737776.

Sekaquaptewa, E., and Washburn, D. (2004). They go along singing: Reconstructing the Hopi past from ritual metaphors in song and image. *American Antiquity, 69*(3), 457–486. https://doi.org/10.2307/4128402.

Wall, R.S. (1989). *A Wild Coast and Lonely: Big Sur Pioneers*. San Carlos, CA: Wide World Publishing/Tetra.

Zagoskin, L.A. (1967). *Lieutenant Zagoskin's Travels in Russian America, 1842–1844* (H.N. Michael, Ed. & P. Rainey, Trans.). University of Toronto Press.

## Chapter 12


Anderson, D. (2000). *Blues for Unemployed Secret Police*. Willimantic, CT: Curbstone Press.

Anderson, D. (2022, May 8). #POETRY: Turning the garden of language. https://www.joysgrape999.com/post/poetry-turning-the-garden-of-language.

Battiata, M. (2003, August 3). Crows: A murder mystery. *The Washington Post Magazine*.

Berreby, D. (2005, September 4). Deceit of the raven. *The New York Times Magazine*. https://www.nytimes.com/2005/09/04/magazine/deceit-of-the-raven.html.

Cao, W. (2015, September 24). Crows (H. Wang, Trans.). *Paper Republic*. https://paper-republic.org/pubs/read/crows/.

Erdrich, L. (2005). *The Painted Drum*. New York: HarperCollins.

Fukase, M. (2017). *Ravens*. London: MACK books.

Hasegawa, A. (2017). Ravens. In *Ravens* (pp.133–136). London: MACK books.

Kilham, L. (1989). *The American Crow and the Common Raven*. College Station: Texas A&M University Press.

Kirpluk, B. (2005). *Caw of the Wild: Observations from the Secret World of Crows*. New York: iUniverse, Inc.

Kosuga, T. (2017). Solitude. In *Ravens* (pp.137–145). London: MACK books.

Lopez, B. (2021). *Desert Notes and River Notes*. New York: Open Road Integrated Media.

"Louis Untermeyer." n.d. Poetry Foundation. https://www.poetryfoundation.org/poets/louis-untermeyer.

Macfarlane, R. (2005, April 2). Seeing the light. *The Guardian*. https://www.theguardian.com/books/2005/apr/02/featuresreviews.guardianreview35.

Marzluff, J., and Angell, T. (2012). *Gifts of the Crow: How Perception, Emotion, and Thought Allow Smart Birds to Behave Like Humans*. New York: Free Press.

Morris, W. (n.d.). Quotes from the artist. *William Morris Studio*. https://www.wmorris.com/artist/quotes-statement/.

Richler, M. (1991). *Solomon Gursky Was Here*. London, England: Vintage Books.

Roethke, T.H. (1975). *The Collected Poems of Theodore Roethke*. New York: Anchor Press.

Skobie, I. (2010, June 28). Lone wolf: An interview with Jim Dine. *Artnet Magazine*. http://www.artnet.com/magazineus/features/scobie/jim-dine6-28-10.asp.

Souter, A., and Archino, S. (2023). Jim Dine: American painter, printmaker, sculptor, poet, conceptual and performance artist. *The Art Story*. https://www.theartstory.org/artist/dine-jim/.

Tibble, J.W. (Ed.). (1935). *The Poems of John Clare* (Vol. II). London: J.M. Dent & Sons, Ltd.

Westerfield, M. (2011). *The Language of Crows*. Willimantic, CT: Ashford Press.

Yardley, J. (1990, May 2). 'Gursky' Richler's national epic. *The Washington Post*. https://www.washingtonpost.com/archive/lifestyle/1990/05/02/gursky-richlers-national-epic/8b6a737d-0517-47a8-8257-74d2b1a8df29/.


## Epilogue


Jensen, A. (1980). A Structural Approach to the Tsimshian Raven Myths: Lévi-Strauss on the Beach. *Anthropologica, 22*(2), 159–186. https://doi.org/10.2307/25605046).

Marzluff, J., and Angell, T. (2012). *Gifts of the Crow: How Perception, Emotion, and Thought Allow Smart Birds to Behave Like Humans*. New York: Free Press.

# Index

Numbers in ***bold italics*** indicate pages with illustrations

www.ingramcontent.com/pod-product-compliance
Lightning Source LLC
LaVergne TN
LVHW081258100826
845148LV00005B/906

* 9 7 8 1 4 7 6 6 9 5 0 0 6 *